CONGRESSIONAL POLICIES, PRACTICES AND PROCEDURES

THE CLEAN AIR ACT AND THE CLEAN POWER PLAN

CONGRESSIONAL POLICIES, PRACTICES AND PROCEDURES

Additional books and e-books in this series can be found on Nova's website under the Series tab.

CONGRESSIONAL POLICIES, PRACTICES AND PROCEDURES

THE CLEAN AIR ACT AND THE CLEAN POWER PLAN

CLAIRE FERGUSON
EDITOR

snova
New York

Library of Congress Cataloging-in-Publication Data

ISBN: 978-1-53614-825-1

Published by Nova Science Publishers, Inc. † New York

CONTENTS

PREFACE

The principal statute addressing air quality concerns, the Clean Air Act was first enacted in 1955, with major revisions in 1970, 1977, and 1990 and is addressed in the first part of this book. Congressional actions on air quality issues have been dominated since 2011 by efforts— particularly in the House—to change the Environmental Protection Agency's (EPA's) authority to promulgate or implement new emission control requirements. EPA's regulations on greenhouse gas emissions from electric power plants and from oil and gas industry sources have been of particular interest, as have the agency's efforts to revise ambient air quality standards for ozone. The 115th Congress and the Trump Administration are reviewing some of these regulations, with the possibility of their modification or repeal. On October 23, 2015, the U.S. Environmental Protection Agency (EPA) published its final Clean Power Plan rule (Rule) to regulate emissions of greenhouse gases (GHGs), specifically carbon dioxide (CO_2), from existing fossil fuel-fired power plants. The aim of the Rule, according to EPA, is to help protect human health and the environment from the impacts of climate change. The Clean Power Plan would require states to submit plans to achieve state-specific CO_2 goals reflecting emission performance rates or emission levels for predominantly coal- and gas-fired power plants, with a series of interim goals culminating in final goals by 2030.

Chapter 1 - Review of regulations issued under the Obama Administration, with the possibility of their modification or repeal, has been the main focus of interest on Clean Air Act issues in the 115th Congress and in the executive and judicial branches in 2017 and 2018. Of particular interest are: the ambient air quality standards for ozone promulgated by the Environmental Protection Agency (EPA) in October 2015; EPA rules to regulate greenhouse gas (GHG) emissions from power plants, cars and trucks, and the oil and gas sector; and emission standards for three groups of sources: brick kilns, wood stoves and heaters, and power plants that burn waste coal. Reducing GHG emissions to address climate change was a major goal of President Obama, but, for a variety of reasons, many in Congress have been opposed to it. In the absence of congressional action, President

Obama directed EPA to promulgate GHG emission standards using existing Clean Air Act authority. This authority has been upheld on three occasions by the Supreme Court, but the specifics remain controversial and subject to debate in Congress, the Trump Administration, and the courts. The Clean Power Plan (CPP), which would limit GHG emissions from existing fossil-fueled power plants, has been a frequent subject of debate. Implementation of the CPP has been stayed by the Supreme Court since February 2016, pending the completion of judicial review. Prior to the stay, challenges to the rule had been filed with the U.S. Court of Appeals for the D.C. Circuit by more than 100 parties, including 27 states. The D.C. Circuit heard oral argument in the case in September 2016; as of this writing, the court has not issued a decision. EPA, under new leadership, has proposed to repeal the CPP and has asked for public input on what should replace it. The D.C. Circuit has stayed the litigation during EPA's review of the CPP. More broadly, on March 28, 2017, President Trump signed Executive Order 13783, to require the review of regulations and policies that burden the development or use of domestically produced energy. The E.O. directed EPA to review the Clean Power Plan and several other regulations for consistency with policies that the E.O. enumerates, and as soon as practicable, to "suspend, revise, or rescind the guidance, or publish for notice and comment proposed rules suspending, revising, or rescinding those rules." GHG rules for methane emissions from the oil and gas industry and from cars and trucks, in addition to the CPP, are subject to the executive order and are under review at EPA, as well as being challenged in the courts. Congress could, of course, short-circuit the EPA and judicial processes, through legislation overturning or modifying any of these regulations. The threat of a filibuster, requiring 60 votes to proceed, might prevent Senate action, however. An EPA rule that has been the subject of action in the 115th Congress is the National Ambient Air Quality Standard (NAAQS) for ozone, which EPA revised in October 2015. Under the revised NAAQS, EPA was to have designated areas that have not attained the standard in October 2017, setting in motion tighter emission requirements for a multitude of industrial, commercial, and mobile sources of pollution. As of March 2018, nonattainment areas have not been identified. On July 18, 2017, the House passed H.R. 806, which would delay designation of the nonattainment areas until 2025 and require future reviews of the NAAQS every 10 years instead of every 5, among other provisions. The House has also passed legislation to delay emission standards for brick kilns and wood heaters (both in H.R. 1917) and to exempt coal-refuse-fired electric generating units (EGUs) from certain emission standards that apply to other coal-fired EGUs (H.R. 1119). All of the House-passed bills await action in the Senate.

Chapter 2 - The Clean Air Act (CAA) gives the Environmental Protection Agency (EPA) broad authority to set ambient air quality standards to protect public health and welfare. It authorizes emission standards for both mobile and stationary air pollution sources, including cars, trucks, factories, power plants, fuels, consumer products, and dozens of other source categories. Since 1970, EPA has used this authority to require emission controls for these sources. Emissions of the most widespread ("criteria") pollutants have been reduced by 72% during that period. As directed by Congress and by executive orders, EPA has estimated the costs and benefits of major CAA (and other) regulations for the last four decades. Its most comprehensive recent studies and studies by the Office of Management and Budget (OMB) have concluded that the benefits of clean air regulations outweigh the costs by substantial margins. EPA's cost-benefit analyses of individual regulations, required by Executive Order 12866, show similar results: a review of the 55 economically significant CAA regulations promulgated from 2001 to 2016 found only two in which estimated costs exceeded benefits. Nevertheless, many in Congress have expressed concern that Clean Air Act and other environmental regulations harm the nation's economy. One issue raised by critics is whether EPA underestimates the cost and other negative impacts of CAA rules—in part, by considering them individually, and not considering cumulative impacts. Another criticism is that the agency relies for most of its benefit assessments on the effects of reducing a single category of pollutants, particulate matter (PM). Research has tied PM to tens of thousands of premature deaths, and EPA often finds that reductions in PM emissions justify regulation, even where PM reductions are a "co-benefit" of reducing another targeted pollutant. A third issue critics raise is whether the methodology used to place monetary value on the avoidance of premature death—a technique referred to as calculating the "value of a statistical life"—inflates the estimated benefits of regulation.

This chapter examines these issues in the context of Clean Air Act regulation. It reviews EPA and Office of Management and Budget (OMB) studies of the cost and benefit of CAA regulations, and addresses the issues raised by agency critics. The chapter finds that

- The Clean Air Act authorizes EPA to set standards in multiple sections of the act: about half of the act's major regulatory authorities mention costs or economic considerations explicitly, and several others imply that costs may be considered; but other authorizing sections, including some key sections, make no mention of cost considerations.
- Where the statutory authorities do not mention cost consideration, they tend to fall into one of four categories: provisions in which Congress itself set the standards; provisions where Congress directed the agency to set health-based standards, without mentioning cost; broad authority to promulgate regulations to achieve an objective that Congress determined was necessary, but the specifics of which it could not anticipate; or authority to promulgate federal requirements in cases where states have failed to develop or implement adequate regulations on their own to meet a federal mandate.

- In all cases, even where the statute would prohibit consideration of cost in setting standards, EPA is bound by executive orders to provide estimates of costs and benefits if the rule would be economically significant.
- According to EPA, the estimated benefits of CAA regulation will exceed the estimated costs by more than 30 to 1 in the period 1990-2020. CAA regulations prevent 230,000 premature deaths annually, according to the agency.
- The estimated benefits of CAA regulations rely heavily on the effects of reducing particulate emissions, and on the value placed on the avoidance of premature death as a result of such controls.
- Many rules have benefits or costs that cannot be quantified or monetized in light of existing information.

President Trump has issued two executive orders that address the cost of EPA regulations: Executive Order (E.O.) 13771, signed January 30, 2017, and E.O. 13783, signed March 28, 2017. The former directs OMB to set regulatory "budgets" for executive branch departments and agencies and, in general, to rescind two regulations for every new one issued. The latter requires EPA to review—and, if appropriate, suspend, revise, or rescind—several CAA regulations affecting energy production, with an eye to avoiding regulatory burdens. At present, the effect of the two orders on future CAA regulations is unclear. The report discusses some of the possible implications.

Chapter 3 - The 1990 Clean Air Act (CAA) amendments required major industrial sources of air pollutants to obtain operating permits. These permits, authorized in Title V of the act, are intended to enhance environmental compliance by detailing for each covered facility all of the emission control requirements to which it is subject. Title V also was intended to generate permit fees that would be used by state and local permitting authorities for administering the program. Implementation of these requirements affects more than 15,000 industrial sources of air emissions, as well as state and local air pollution control agencies. Adding these provisions to the act was controversial, and implementation, too, has generated controversies. The Environmental Protection Agency (EPA) issued regulations to implement Title V in 1992. Aspects of those rules (particularly concerning procedures to modify permits) have been contentious since then. EPA has considered a number of regulatory revisions but has not finalized any modifications. However, EPA has issued white papers and a number of formal and informal guidance documents that, together with the 1992 rules, comprise the agency's current interpretation of statutory and regulatory requirements. Because of regulatory and program approval delays, state and local agencies were slow to begin issuing Title V permits, falling far short of statutory deadlines and EPA's goals. According to an EPA Inspector General report, key factors that delayed issuance of permits included insufficient state resources, complex EPA rules and limited guidance, and conflicting state priorities. Now, however, most initial permits have been issued, and permit reissuance (required after five years) and modification have replaced issuance of initial permits as the major ongoing task of permitting agencies. Attention to the Title V program increased in 2010 when EPA initiated several controversial regulatory actions to regulate emissions of greenhouse gases (GHGs) under existing CAA authority with implications for Title V permits. For Title V, these actions mean including GHG control requirements in Title V permits issued for non-GHG. To minimize the costs and administrative burden of its GHG regulations, EPA issued a "Tailoring Rule" to impose requirements only on the largest sources of GHG emissions. In June 2014, the Supreme Court found that EPA exceeded its statutory authority in issuing the Tailoring Rule but upheld the agency's authority to require sources that already need permits for conventional pollutant emissions to comply with CAA requirements for GHGs. Most stakeholders agree that at least some of the benefits of Title V identified by Congress in the 1990 CAA amendments have been achieved, such as incorporation of applicable air pollution control requirements in a single document that is accessible to regulators, the public, and industrial sources. At the same time, there also has been widespread dissatisfaction with the program's complexity, costs, and confusing requirements. Many believe that a lack of EPA guidance and oversight has contributed greatly to implementation problems. Congressional examination of Title V has been limited to a few oversight hearings, but none recently. Clean Air Act issues have been of considerable interest during the 114th Congress, especially scrutiny of EPA's regulation of greenhouse gas emissions. Although EPA's actions concerning GHGs involve multiple provisions of the act, this congressional attention has not included Title V.

Chapter 4 - This chapter provides a selective overview of court decisions that historically have most shaped EPA's program under the Clean Air Act (CAA or Act). Court decisions described in the chapter deal with the following:

- *National ambient air quality standards (NAAQSs),* holding that in setting the standards EPA is not to consider economic and technological feasibility.
- *State implementation plans* for achieving NAAQSs, holding that EPA may not consider economic and technological feasibility in approving or disapproving such plans, or that the state plan is more stringent than necessary, or does not require an EPA-preferred control method.
- *Interstate air pollution*, holding that EPA may consider costs in applying the CAA "good neighbor" provision, but any emissions trading program must assure some emission reduction in each upwind state. Nor does the CAA require that states be given a second opportunity to file an implementation plan after EPA has quantified the state emissions budget; EPA may promulgate its own plan for the state immediately.
- *New source performance standards (NSPSs),* holding that while the Act requires NSPSs to be based on "adequately demonstrated" technology, EPA is allowed to

consider technologies that will be fairly projected in the future so long as the technology is not speculative.

- *New source review* in areas not subject to NAAQSs, holding that EPA may override a state's determination of the "best available control technology" required for new stationary sources. EPA may require new source review for greenhouse gas emitters only if the new source will emit certain pollutants above threshold amounts.
- *The "routine maintenance" exemption* from NSPSs and new source review, created by EPA and accepted by the courts despite statutory silence. Courts hold that whether the exemption applies depends on the increase in a plant's expected life due to the project, and the project's cost, nature, and magnitude. Expansive interpretation of the exemption has been judicially rejected.
- *The "bubble concept,"* an EPA approach that looks at *net* changes in the emissions of a pollutant from a facility, holding that its permissibility depends on statutory context.
- *National standards for hazardous air pollutants*, holding that EPA may determine if a facility triggers the Act's "maximum achievable control technology" requirement for such pollutants by aggregating emission sources in a contiguous plant under common control, not just sources within the same source category. EPA is not limited in setting emission standards to hazardous air pollutants currently controlled with technology.
- *Greenhouse gas emissions*, holding that the CAA generally covers them, and that EPA has to exercise that authority based on policy concerns. See, however, "new source review" above.
- *Enforcement*, holding that the recipient of an administrative compliance order must be allowed to seek pre-enforcement review of the order in court.

Chapter 5 - Notwithstanding air quality progress since 1970, challenges remain to reduce pollution in areas exceeding federal standards and to ensure continued compliance elsewhere. The movement of air pollutants across state lines, known as interstate transport, has made it difficult for some downwind states to attain federal ozone and fine particulate matter ($PM_{2.5}$) standards, partly because states lack authority to limit emissions from other states. The Clean Air Act's "Good Neighbor" provision (Section 110(a)(2)(D)) seeks to address this issue and requires states to prohibit emissions that significantly contribute to another state's air quality problems. It requires each state's implementation plan (SIP)—a collection of air quality regulations and documents—to prohibit emissions that either "significantly contribute" to nonattainment or "interfere with maintenance" of federal air quality standards in another state. The act also authorizes states to petition EPA to issue a finding that emissions from "any major source or group of stationary sources" violate the Good Neighbor provision (Section 126(b)).

EPA and the states have implemented regional programs to address interstate ozone and $PM_{2.5}$ transport and comply with the Good Neighbor provision. These programs set emission "budgets" for ozone and $PM_{2.5}$ precursor emissions—specifically, sulfur dioxide (SO_2) and nitrogen oxide (NO_x) as $PM_{2.5}$ precursors and seasonal NO_x emissions as an ozone precursor. The current program—the Cross State Air Pollution Rule (CSAPR)—focuses on limiting interstate transport of power sector SO_2 and NO_x emissions to eastern states. Power sector emissions in CSAPR states are below emission budgets as a result of regulatory and market factors (see figure). Annual SO_2, annual NO_x, and ozone season NO_x emissions from CSAPR sources decreased 77%, 41%, and 15%, respectively, between 2009 and 2016. EPA has concluded that regional SO_2 and NO_x programs have reduced interstate transport of $PM_{2.5}$ and ozone. The Energy Information Administration's national-scale analysis identifies market and regulatory factors contributing to emission reductions. It is unclear whether emissions will remain well below budgets, given recent prices of ozone season NO_x allowances (i.e., authorization for each ton emitted) and the supply of banked allowances for future use in lieu of emission reductions. Research indicates that ozone transport harms air quality in downwind states. However, stakeholder views vary regarding the extent to which interstate transport impacts air quality. Some note that some coal-fired power plants do not fully use already-installed pollution controls. Several states have sought additional upwind reductions in ozone precursors through Section 126(b) petitions. Others have questioned the feasibility of achieving additional reductions in ozone precursors, raising concerns about emissions from international or natural sources. EPA recently proposed to determine that CSAPR fully addresses Good Neighbor obligations for the 2008 ozone standard but has not yet made a "Good Neighbor" determination for the more stringent 2015 ozone standard. The agency has therefore not yet determined whether and how it will update the CSAPR budgets with respect to the 2015 ozone standard. Members of Congress may have an interest in better understanding how EPA and states implement the Clean Air Act's Good Neighbor provision, particularly as EPA continues its assessment of Good Neighbor obligations under the 2015 ozone standard. The following issues, among others, may inform deliberations about interstate air transport. First, the extent to which existing programs will improve air quality in areas not meeting the 2015 ozone standard is to be determined. CSAPR has not addressed NO_x emissions from nonpower sector sources, such as large industrial boilers. EPA concluded that industrial sources have potential to cost-effectively reduce NO_x emissions but is less certain about the structure of potential NO_x control strategies. Second, some have questioned whether additional regulatory incentives are necessary to fulfill Good Neighbor obligations, particularly given current NO_x allowance prices. These prices are below the marginal abatement cost, which may result in higher emissions. Third, EPA's current air quality initiatives may indirectly affect its Good Neighbor assessments. EPA recently sought comment on potential flexibilities for the development of Good Neighbor SIPs.

technological innovation, and federal tax incentives. Many coal-fired power plants are being retired, and the new electric generation replacing those plants is overwhelmingly powered by natural gas or renewable power. Because coal-fired plants emit far more CO_2 per unit of power than their replacements, total emissions of CO_2 from electric power generation declined almost 25% between 2005 and 2016, while gross domestic product grew and the amount of power generated remained essentially unchanged. This observed decline in annual CO_2 emissions from the electric power sector is 77% of the reductions that EPA projected would occur as a result of the CPP. Members of Congress may have an interest—for legislative and oversight purposes, as potential commenters, and in responding to constituents—in understanding what it is that EPA has proposed to do with regard to the CPP. This chapter provides background on the CPP and its proposed repeal, describes the administrative steps that are required to repeal or amend a rule, and discusses how the CPP and its proposed repeal fit into the context of recent and projected power sector evolution.

Chapter 9 - On October 23, 2015, the U.S. Environmental Protection Agency (EPA) published its final Clean Power Plan rule (Rule) to regulate emissions of greenhouse gases (GHGs), specifically carbon dioxide (CO2), from existing fossil fuel-fired power plants. The aim of the Rule, according to EPA, is to help protect human health and the environment from the impacts of climate change. The Clean Power Plan would require states to submit plans to achieve state-specific CO_2 goals reflecting emission performance rates or emission levels for predominantly coal- and gas-fired power plants, with a series of interim goals culminating in final goals by 2030. The Clean Power Plan has been one of the more singularly controversial environmental regulations ever promulgated by EPA, and the controversy is reflected in the enormous multiparty litigation over the Rule ongoing in the U.S. Court of Appeals for the District of Columbia Circuit (D.C. Circuit). Numerous petitions challenging the Clean Power Plan have been consolidated into one case, *West Virginia v. EPA*. While the litigation is still ongoing at the circuit court level, an unusual mid-litigation application to the Supreme Court resulted in a stay of the Rule, meaning that the Rule does not have legal effect at least for the duration of the litigation. On September 27, 2016, the en banc (full court) D.C. Circuit heard oral argument for the case. This chapter provides legal background on the Rule, its Clean Air Act (CAA) framework under Section 111, and climate-related lawsuits that have preceded the present litigation over the Clean Power Plan. It then gives an overview of the participants in the current litigation, including two groups of Members of Congress, who have offered briefs in support of both sides. This chapter highlights the major events in the litigation as of the date of publication, including the Supreme Court stay and oral argument, and the likely timetable of events in the near term. Some of the main arguments on the merits are then briefly summarized and excerpted from court filings, including

- the standard of review to apply to EPA's action;
- the scope of EPA's overall authority under CAA Section 111;
- whether Section 111 allows the Clean Power Plan's inclusion of generation-shifting, such as from coal-fired power plants to lower-emitting sources of electricity;
- the interpretation of a statutory exclusion in CAA Section 111 that cross-references CAA Section 112's regulation of hazardous air pollutants, particularly in light of the apparent enactment in 1990 of differing House and Senate amendments to the same cross-reference;
- constitutional arguments relating to federalism and separation of powers;
- record-based challenges to the achievability and reasonableness of the Rule; and
- arguments regarding rulemaking procedures.

This chapter concludes with a brief look at parallel litigation in the D.C. Circuit, consolidated as *North Dakota v. EPA*, which is challenging a related EPA regulation that imposes new source performance standards (NSPSs) limiting CO_2 emissions from new, modified, and reconstructed fossil fuel-fired power plants.

Chapter 10 - On December 28, 2017, the U.S. Environmental Protection Agency (EPA) published an Advanced Notice of Proposed Rulemaking (ANPRM) seeking information on a potential replacement to the Clean Power Plan (CPP)—the emissions guidelines for greenhouse gases (GHGs) from existing power plants that the agency proposed to repeal in October 2017. An ANPRM allows an agency to gather information for potential rulemaking but does not bind an agency to future action. As explained in the ANPRM, EPA is "considering" proposing a rule to replace the CPP and seeks comments on the (1) "roles, responsibilities, and limitations of the federal government, state governments, and regulated entities in developing and implementing" the new GHG emission guidelines; (2) the "best system of emission reduction" consistent with EPA's new legal interpretation set forth in the proposed repeal of the CPP; and (3) approaches to harmonize and streamline the process and applicability of Clean Air Act permitting for existing power plants regulated under the new emission guidelines. EPA will accept comments on the ANPRM until February 26, 2018. This Legal Sidebar post originally appeared November 13, 2017. On October 10, 2017, the Environmental Protection Agency (EPA) issued a proposal to repeal the Obama Administration's 2015 rule, "Carbon Pollution Emission Guidelines for Existing Stationary Sources: Electric Utility Generating Units (EGUs)" (commonly referred to as the Clean Power Plan (CPP) rule). The CPP has not gone into effect because the Supreme Court in 2016 stayed the implementation of the rule until the lawsuit challenging its legality is resolved. Upon its review of the CPP and its 2015 legal justification, EPA has now determined that the CPP exceeds its statutory authority based on a change in the agency's legal interpretation of Section 111 of the Clean Air Act (CAA). The proposal formally starts a potentially lengthy process to repeal the CPP and raises questions about whether EPA will replace the CPP with another rule targeting CO_2 emissions from existing power plants and how the repeal will affect existing legal challenges to the CPP. This Sidebar will explore these and other questions and the next steps in repealing the CPP.

Chapter 11 - Recent international negotiations and domestic policy developments have generated interest in current and projected U.S. greenhouse gas (GHG) emission levels. GHG emissions are generated throughout the United States from millions of discrete sources. Of the GHG source categories, carbon dioxide (CO_2) emissions from fossil fuel combustion account for the largest percentage (77%) of total U.S. GHG emissions. The electric power sector contributes the second largest percentage (35%) of CO_2 emissions from fossil fuel combustion (one percentage point behind the transportation sector). In December 2015, delegations from 195 nations, including the United States, adopted an agreement in Paris that creates an international structure for nations to pledge to abate their GHG emissions, adapt to climate change, and cooperate to achieve these ends, including financial and other support. Pursuant to that agreement, the United States pledged (in 2015) to reduce GHG emissions by 26-28% by 2025 compared to 2005 levels. At the date of this chapter, U.S. involvement in the Paris Agreement remains uncertain. However, some recent reports indicate that President Trump is expected to withdraw from the agreement. U.S. GHG levels in 2015 were 11% below 2005 levels. Whether the United States achieves its goals would likely depend, to some degree, on CO_2 emissions from power plants. In 2015, under President Obama, the Environmental Protection Agency (EPA) promulgated standards for CO_2 emissions from existing electric power plants. The rule, known as the Clean Power Plan (CPP), is the subject of ongoing litigation involving a number of entities. On February 9, 2016, the Supreme Court stayed the rule for the duration of the litigation. Multiple factors generally impact CO_2 emission levels from the electric power sector. Some factors are listed below in no particular order:

- Economic growth/recession,
- Relative prices of energy sources for electricity—particularly natural gas and renewable energy sources,
- Electricity generation portfolio (i.e., the ratio of electricity generation from coal, natural gas, and renewable energy sources),
- National and/or state policy developments (e.g., CPP implementation), and
- Demand-side efficiency improvements (e.g., commercial and residential electricity use).

Recent changes in the electric power sector may be informative. Between 1975 and 2010, electricity generation and CO_2 emissions from the electric power sector generally increased. However, in 2010, their courses diverged. While electricity generation remained relatively flat after 2010, CO_2 emissions from the electric power sector decreased. Thus in

2016, electricity generation was essentially equivalent to generation in 2005, while CO_2 emissions were 25% below 2005 levels.

Recent changes in the U.S. electricity generation portfolio played a key role in the CO_2 emission decrease. The electricity portfolio affects CO_2 emission levels because different sources of electricity generation produce different rates of CO_2 emissions per unit of electricity (zero in the case of some renewables). For example, between 2005 and 2016:

- Coal's contribution to total electricity generation decreased from 50% to 30%;
- Natural gas's contribution to total electricity generation increased from 19% to 34%; and
- Renewable energy's contribution to total electricity generation increased from 2% to 8%.

Accurately forecasting future CO_2 emission levels is a complex and challenging endeavor. A comparison of actual CO_2 emissions (from energy use) between 1990 and 2017 with selected emission projections illustrates this difficulty. In general, actual emissions have remained well below projections. As the future of the CPP is uncertain, some have questioned whether existing policies and trends in electricity generation would continue to lower CO_2 emissions. Modeling results indicate that CO_2 emissions in the electricity sector are expected to continue declining. However, modeling results indicate that the declines would be substantially greater if the CPP were implemented.

PART 1: CLEAN AIR ACT

In: The Clean Air Act and The Clean Power Plan
Editor: Claire Ferguson
ISBN: 978-1-53614-825-1

Chapter 1

CLEAN AIR ACT ISSUES IN THE 115TH CONGRESS: IN BRIEF*

James E. McCarthy

SUMMARY

Review of regulations issued under the Obama Administration, with the possibility of their modification or repeal, has been the main focus of interest on Clean Air Act issues in the 115th Congress and in the executive and judicial branches in 2017 and 2018. Of particular interest are: the ambient air quality standards for ozone promulgated by the Environmental Protection Agency (EPA) in October 2015; EPA rules to regulate greenhouse gas (GHG) emissions from power plants, cars and trucks, and the oil and gas sector; and emission standards for three groups of sources: brick kilns, wood stoves and heaters, and power plants that burn waste coal.

Reducing GHG emissions to address climate change was a major goal of President Obama, but, for a variety of reasons, many in Congress have been opposed to it. In the absence of congressional action, President Obama directed EPA to promulgate GHG emission standards using existing Clean Air Act authority. This authority has been upheld on three occasions by the Supreme Court, but the specifics remain controversial and subject to debate in Congress, the Trump Administration, and the courts.

The Clean Power Plan (CPP), which would limit GHG emissions from existing fossil-fueled power plants, has been a frequent subject of debate. Implementation of the CPP has been stayed by the Supreme Court since February 2016, pending the completion of judicial review. Prior to the stay, challenges to the rule had been filed with the U.S. Court of Appeals for the D.C. Circuit by more than 100 parties, including 27 states. The D.C. Circuit heard oral argument in the case in September 2016; as of this writing, the court has not issued a decision. EPA, under new leadership, has proposed to repeal the CPP and has

* This is an edited, reformatted and augmented version of a Congressional Research Service report, Publication No. R44744, dated March 12, 2018.

supported national standards, in part, to avoid having to meet standards on a state-by-state basis.

The second round of GHG standards for cars and light trucks is being phased in over model years (MY) 2017-2025. As part of the rulemaking, EPA made a commitment to conduct a Mid-Term Evaluation (MTE) for the MY2022-2025 standards by April 2018. The agency deemed an MTE appropriate given the long time frame at issue, with the final standards taking effect as long as 12 years after promulgation. Through the MTE, EPA was to determine whether the standards for MY2022-2025 were still appropriate given the latest available data and information, with the option of strengthening, weakening, or retaining the standards as promulgated.

On November 30, 2016, EPA released a proposed determination under the MTE stating that the MY2022-2025 standards remained appropriate and that a rulemaking to change them was not warranted. EPA based its findings on a Technical Support Document, a previously released Draft Technical Assessment Report (which was issued jointly by EPA, DOT, and the California Air Resources Board [CARB]), and input from the auto industry and other stakeholders. The proposed determination opened a public comment period that ran through December 30, 2016. On January 12, 2017, the EPA Administrator made a final determination to retain the MY2022- 2025 standards as originally promulgated.

This action significantly accelerated the original timeline for the MTE (which called for a final determination by April 2018), and EPA announced it separately from any DOT (fuel economy) or California (GHG standard) process. (California subsequently reaffirmed its commitment to the 2022-2025 standards, on March 24, 2017.) Critics reacted to the accelerated timetable swiftly, vowing to work with the Trump Administration to revisit EPA's determination—citing a "rush to judgment" that they argued contradicted the objectives of the One National Program. Among the potential revisions suggested by critics has been better harmonization of the existing EPA/DOT/CARB standards, easing the MY2022-2025 standards, or eliminating them entirely.

The Administration reopened the MTE in mid-March 2017, but it is unclear to what extent EPA will modify or eliminate the standards, or if so, how quickly. To change the existing rule would require a new rulemaking, a process that normally would take a year or more and would be subject to judicial and congressional review upon completion. A rulemaking would probably need to be justified by new data or a reasonable justification to avoid being considered arbitrary and capricious if challenged in court. Some changes, on the other hand, such as better coordination of the standards with those of DOT and California, might be implemented without a change to the promulgated EPA standards. With the options for review uncertain, the role of Congress may be one of oversight, at least for now.

EPA and DOT have also promulgated joint GHG emission and fuel economy standards for medium- and heavy-duty trucks,[4] which have generally been supported by the trucking industry and truck and engine manufacturers. This rule was finalized on August 16, 2016.[5] The new standards cover MY2018-2027 for certain trailers and model years MY2021-2027 for semi-trucks, large pickup trucks, vans, and all types and sizes of buses and work trucks. According to EPA,

> The Phase 2 standards are expected to lower CO_2 emissions by approximately 1.1 billion metric tons, save vehicle owners fuel costs of about $170 billion, and reduce oil consumption by up to 2 billion barrels over the lifetime of the vehicles sold under the program.[6]

In the Regulatory Impact Analysis accompanying the rule's promulgation, EPA projected the total cost of the Phase 2 standards at $29-$31 billion over the lifetime of MY2018-2029 trucks. The standards will increase the cost of a long haul tractor-trailer by as much as $13,500 in MY2027, according to the agency; but the buyer would recoup the investment in fuel-efficient technology in less than two years through fuel savings. In EPA's analysis, fuel consumption of 2027 model tractor-trailers will decline by 34% as a result of the rule.[7]

In general, the truck standards have been well-received. The American Trucking Associations, for example, described themselves as "cautiously optimistic" that the rule would achieve its targets: "We are pleased that our concerns such as adequate lead-time for technology development, national harmonization of standards, and flexibility for manufacturers have been heard and included in the final rule."[8] The Truck and Engine Manufacturers Association, while describing itself as "in the process of reviewing" the final rule, highlighted its work providing input to assure that EPA and DOT established a single national program, and concluded: "A vitally important outcome is that EPA and

---

[4] U.S. Environmental Protection Agency, U.S. Department of Transportation, "Greenhouse Gas Emissions Standards and Fuel Efficiency Standards for Medium- and Heavy-Duty Engines and Vehicles; Final Rules," 76 *Federal Register* 57106, September 15, 2011.

[5] The rule appeared in the *Federal Register* on October 25, 2016: U.S. Environmental Protection Agency, U.S. Department of Transportation, "Greenhouse Gas Emissions and Fuel Efficiency Standards for Medium- and Heavy-Duty Engines and Vehicles—Phase 2; Final Rule," 81 *Federal Register* 73478, 73482, October 25, 2016. Fact sheets and links to the final rule and the Regulatory Impact Analysis are at https://www.epa.gov/regulations-emissionsvehicles-and-engines/final-rule-greenhouse-gas-emissions-and-fuel-efficiency.

[6] U.S. Environmental Protection Agency, Office of Transportation and Air Quality, "EPA and NHTSA Adopt Standards to Reduce Greenhouse Gas Emissions and Improve Fuel Efficiency of Medium- and Heavy-Duty Vehicles for Model Year 2018 and Beyond; Regulatory Announcement," August 2016, at https://nepis.epa.gov/Exe/ZyPDF.cgi/ P100P7NL.PDF?Dockey=P100P7NL.PDF.

[7] Ibid.

[8] American Trucking Associations, "ATA Hopes Final Truck Efficiency Rule Will Achieve Emissions Goals," Press Release, August 16, 2016, at http://www.trucking.org/article/ATA-Hopes-Final-Truck-Efficiency-Rule-Will-AchieveEmissions-Goals.

DOT have collaborated to issue a single final rule that includes a harmonized approach to greenhouse gas reductions and fuel efficiency improvements."[9]

Neither group filed petitions for judicial review of the rule. The only challengers were the Truck Trailer Manufacturers Association and the Racing Enthusiasts and Suppliers Coalition. Nevertheless, in April 2017, EPA took steps to review the rule. The agency asked the D.C. Circuit Court of Appeals to hold the legal challenge (*Truck Trailer Manufacturers Association v. EPA*) in abeyance while it conducts a review of the standards. The court granted EPA's request on May 8, 2017. On October 27, 2017, the D.C. Circuit Court granted the Truck Trailer Manufacturers Association's request to stay certain requirements for trailers pending the judicial review of the medium- and heavy-duty vehicles rule.[10] The rest of the rule remains in effect. The truck rule also established emission standards for vehicles manufactured from "glider kits" (truck bodies produced without a new engine, transmission, or rear axle). On November 16, 2017, EPA proposed to repeal the glider kit provisions.

Standards for Power Plants (Clean Power Plan)

Power plants are the largest individual anthropogenic sources of U.S. GHG emissions; as a group, they account for about 30% of the U.S. total. EPA finalized emission standards for new, existing, and modified fossil-fueled power plants in August 2015.[11] The standards would primarily affect coal-fired units, which emit twice the amount of carbon dioxide (CO_2) that would be emitted by an equivalent natural gas combined cycle (NGCC) electric generating unit.[12] The final rules have been hugely controversial: EPA received more than 4 million public comments as it considered the proposed standards for existing units, by far the most comments on a rulemaking in the agency's 46-year history.

The Clean Power Plan (CPP), which is the rule for existing units, would set state-specific goals for CO_2 emissions and emission rates from existing fossil-fueled power plants. EPA established different goals for each state based on three "building blocks": improved efficiency at coal-fired power plants; substitution of NGCC generation for coal-fired power; and zero-emission power generation from increased renewable energy, such as wind or solar. Two sets of goals were established by the rule: an interim set, which would apply to the average emissions rate in a state in the 2022-2029 time period; and a

---

[9] Truck and Engine Manufacturers Association, "Truck and Engine Manufacturers Evaluating New Phase 2 Greenhouse Gas Regulations that EPA/DOT Announced Today," Press Release, August 16, 2016, at http://www.truckandenginemanufacturers.org/file.asp?A=Y&F=2016+08+16+Final++GHG+Phase+II +Press+Release.pdf&N=2016+08+16+Final ++GHG+Phase+II+Press+Release.pdf&C=documents.

[10] *Order, Truck Trailer Manufacturers Ass'n v. EPA*, No. 16-1430 (October 27, 2017).

[11] Links to the Clean Power Plan, as well as extensive background materials, can be found on EPA's website at https://19january2017snapshot.epa.gov/cleanpowerplan/clean-power-plan-existing-power-plants_.html.

[12] See CRS Report R44090, *Life-Cycle Greenhouse Gas Assessment of Coal and Natural Gas in the Power Sector*.

final state-specific average emission rate for the years 2030 and beyond. States can reach these goals through a wide array of options, including heavier reliance on renewable or nuclear power; reductions in power demand through efficiency programs; substituting natural gas-fired for coal-fired units; the use of tradeable allowances; and combining efforts with other states. In general, states that currently rely on coal-fired power to a great extent would be allowed higher emission rates, but would have to reduce average emissions by a greater percentage than other states.[13] The first step in implementation was to be the submission of implementation plans by the states in September 2016.

A separate rule for new power plants, the New Source Performance Standards (NSPS), would affect fewer plants, but it too is controversial, because of the technology it assumed could be used to reduce emissions at new coal-fired units. The NSPS would rely in part on carbon capture and sequestration (CCS) technology to reduce emissions by about 20% compared to the emissions of a state of the art coal-fired plant without CCS. Critics complained that CCS is a costly and unproven technology, and because of this, the NSPS would effectively prohibit the construction of new coal-fired plants. They noted that no operating commercial U.S. power plant was capturing and storing CO_2 as of the date the rule was promulgated. (The first commercial CCS facility in the United States, the Petra Nova project at the W.A. Parish Generating Station in Texas, came on line in 2016.)

In promulgating the rule, EPA contended that the components of CCS (separation of CO_2 from emissions, pipelines to transport the CO_2 to storage sites, and equipment to pump the CO_2 underground) have been successfully operated for decades, even if no power plant had combined all of them in an operating unit. At the same time, EPA maintained that the NSPS rule "will result in negligible ... costs," because, given the low cost and abundance of natural gas, new fossil-fueled capacity will rely on natural gas (NGCC) technology for the immediate future.[14] NGCC units, which emit only half the CO_2 of uncontrolled coal-fired plants, can attain the NSPS without needing to capture any of their carbon emissions.

Following publication of the NSPS and the Clean Power Plan in the October 23, 2015, *Federal Register*, the 114th Congress considered and passed joint resolutions of disapproval of both rules in December 2015 (S.J.Res. 23 and S.J.Res. 24) under the Congressional Review Act (CRA). The CRA resolutions would have revoked the rules and prevented

---

[13] For a more detailed description of the CPP requirements, see CRS Report R44145, *EPA's Clean Power Plan: Highlights of the Final Rule*, and CRS Report R44341, *EPA's Clean Power Plan for Existing Power Plants: Frequently Asked Questions*.

[14] .S. Environmental Protection Agency, Regulatory Impact Analysis for the Final Standards of Performance for Greenhouse Gas Emissions from New, Modified, and Reconstructed Stationary Sources: Electric Utility Generating Units, August 2015, pp. ES-4 and ES-5, at https://nepis.epa.gov/Exe/ZyNET.exe/P100MWLE.txt?ZyActionD=ZyDocument&Client=EPAIndex=2011%20Thru%202015&Docs=&Query=&Time=&EndTime=&SearchMethod=1&TocRestrict=n&Toc=&TocEntry=&QField=&QfieldYear=&QFieldMonth=&QFieldDay=&UseQField=&IntQFieldOp=0&ExtQFieldOp=0&XmlQuery=&File=D%3A%5CZYFILES%5CINDEX%20DATA%5C11THRU15%5CTXT%5C00000016%5CP100MWLE.txt&User=ANONYMOUS&Password=anonymous&SortMethod=h%7C-&MaximumDocuments=1&FuzzyDegree=0&ImageQuality=r75g8/r75g8/x150y150g16/i425&Display=hpfr&DefSeekPage=x&SearchBack=ZyActionL&Back=ZyAction S&BackDesc=Results%20page&MaximumPages=1&ZyEntry=1.

EPA from issuing substantially similar rules unless authorized by a subsequently enacted law. President Obama vetoed both of the joint resolutions.

The CRA resolutions were the latest in a long line of attempts by Members, primarily in the House, to limit EPA's authority to implement GHG emission requirements for power plants (referred to as electric generating units, or EGUs, in EPA parlance). For example, in June 2015, the House passed H.R. 2042, which would have delayed the compliance date of GHG emission standards for existing EGUs (including the date by which states must submit implementation plans) until after the completion of judicial review of any aspect of the rule, and would have allowed a state to opt out of compliance if the governor determined that the rule would have significant adverse effects on rate-payers or on the reliability of the state's electricity system.

Legislation was also considered in the 113th and 112th Congresses. In the 113th, the House passed H.R. 3826, which would have prohibited EPA from promulgating or implementing GHG emission standards for fossil-fueled power plants until at least six power plants representative of the operating characteristics of electric generation units at different locations across the United States had demonstrated compliance with proposed emission limits for a continuous period of 12 months on a commercial basis. Projects demonstrating the feasibility of carbon capture and storage that received government financial assistance could not have been used in setting such standards, and the standards would not have taken effect unless Congress enacted new legislation setting an effective date. The House incorporated the language of H.R. 3826 in H.R. 2, which also passed the House. The House also passed three bills in the 112th Congress. The Senate did not take up any of the House bills, however.

Another frequently discussed option to prevent EPA action on GHG emissions would be an appropriations rider prohibiting EPA from finalizing or implementing the EGU standards. On July 14, 2016, the House passed H.R. 5538, the Department of the Interior, Environment, and Related Agencies Appropriations Act, 2017. Section 431 of the bill would have prevented the use of funds appropriated under the bill to implement or enforce both the NSPS and the Clean Power Plan. The House rider was not included in the continuing resolutions (P.L. 114-223 and P.L. 114-254) that funded EPA through April 2017 or in the Consolidated Appropriations Act that funded EPA through the remainder of the fiscal year (P.L. 115-31).

Action on the CPP was a frequent topic in articles discussing the 115th Congress's priorities at the beginning of 2017, but the courts and EPA have turned out to be the venues for action. Implementation of the CPP was stayed by the Supreme Court in February 2016, pending the completion of judicial review. Challenges to the rule were filed with the U.S. Court of Appeals for the D.C. Circuit by more than 100 parties, including 27 states. These challenges have been consolidated into a single case, *West Virginia v. EPA*. The D.C. Circuit heard oral argument in the case in September 2016; as of this writing, the court has not issued a decision. The NSPS have also been challenged (*North Dakota v. EPA*). EPA requested (and the court granted) a pause in the litigation to give it time to conduct a review.

Under the Trump Administration, EPA now appears the more likely venue to modify the CPP and the NSPS. On March 28, 2017, President Trump signed Executive Order 13783, to require the review of regulations and policies that burden the development or use of domestically produced energy. The E.O. directed EPA to review the Clean Power Plan, the NSPS, and other regulations for consistency with policies that the E.O. enumerates, and as soon as practicable, to "suspend, revise, or rescind the guidance, or publish for notice and comment proposed rules suspending, revising, or rescinding those rules." EPA proposed repeal of the CPP on October 16, 2017,[15] and has issued an Advance Notice of Proposed Rulemaking to solicit information on systems of emission reduction that it might require in a future rule to replace it.[16] Like judicial review, this route could be time-consuming. Repealing or replacing the rule will require the agency to follow the administrative steps involved in proposing and promulgating a new rule. Following promulgation, the repeal action and any replacement rule will themselves be subject to judicial review. A large group of stakeholders, including some states, will likely oppose major changes to the CPP.

The EPA and judicial processes could be short-circuited by Congress, through legislation overturning or modifying the CPP or NSPS. The threat of a filibuster, requiring 60 votes to proceed, might prevent Senate action, however.

Standards for the Oil and Gas Industry

On June 3, 2016, EPA promulgated a suite of New Source Performance Standards (NSPS) under Section 111 of the Clean Air Act to set controls for the first time on methane emissions from sources in the crude oil and natural gas production sector and the natural gas transmission and storage sector.[17] The rule builds on the agency's 2012 NSPS for volatile organic compound (VOC) emissions[18] and would extend controls for methane and

---

[15] U.S. Environmental Protection Agency, Repeal of Carbon Pollution Emission Guidelines for Existing Stationary Sources: Electric Utility Generating Units, 82 Federal Register 48,035, October 16, 2017.

[16] U.S. Environmental Protection Agency, State Guidelines for Greenhouse Gas Emissions from Existing Electric Utility Generating Units, Advance Notice of Proposed Rulemaking, 82 Federal Register, 61507, December 28, 2017. A link to the document and an EPA Fact Sheet can be found at https://www.epa.gov/stationary-sources-air-pollution/electric-utility-generating-units-advance-notice-proposed.

[17] U.S. Environmental Protection Agency, "Oil and Natural Gas Sector: Emission Standards for New, Reconstructed, and Modified Sources, Final Rule," 81 *Federal Register* 35824, June 3, 2016.

[18] U.S. Environmental Protection Agency, "Oil and Natural Gas Sector: New Source Performance Standards and National Emission Standards for Hazardous Air Pollutants Reviews; Final Rule," 77 *Federal Register* 49489, August 16, 2012.

VOC emissions beyond the existing requirements to include new or modified hydraulically fractured oil wells, pneumatic pumps, compressor stations, and leak detection and repair at well sites, gathering and boosting stations, and processing plants. The Administration stated that the rule was a key component under President Obama's "Climate Action Plan," and that the plan's Strategy to Reduce Methane Emissions[19] was needed to set the Administration on track to achieve its goal to cut methane emissions from the oil and gas sector by 40%-45% from 2012 levels by 2025, and to reduce all domestic greenhouse gas emissions by 26%-28% from 2005 levels by 2025.

Methane—the key constituent of natural gas—is a potent greenhouse gas with a global warming potential (GWP) more than 25 times greater than that of carbon dioxide. According to EPA's *Inventory of U.S. Greenhouse Gas Emissions and Sinks*, methane is the second most prevalent GHG emitted in the United States from human activities, and nearly 30% of those emissions come from oil production and the production, transmission, and distribution of natural gas.[20]

EPA projected that the standards for new, reconstructed, and modified sources will reduce methane emissions by 510,000 tons in 2025, the equivalent of reducing 11 million metric tons of carbon dioxide.[21] In conjunction with the proposal, EPA conducted a Regulatory Impact Analysis (RIA) that looked at the illustrative benefits and costs of the proposed NSPS: in 2025, EPA estimated the rule will have costs of $530 million and climate benefits of $690 million (in constant 2012 dollars). The rule will also reduce emissions of VOCs and hazardous air pollutants (HAPs). EPA was not able to quantify the benefits of the VOC/HAP reductions.

The methane rule is among the rules subject to review under E.O. 13783. Section 7 of the E.O. directed EPA to review the rule for consistency with policies that the E.O. enumerates, and, if appropriate, as soon as practicable, to "suspend, revise, or rescind the guidance, or publish for notice and comment proposed rules suspending, revising, or rescinding those rules." EPA has begun a review of the rule and has proposed a two-year stay of the rule during the review process. The agency also issued a 90-day administrative stay of portions of the rule on June 5, 2017. The U.S. Court of Appeals for the D.C. Circuit vacated the 90-day stay on July 3, 2017, however, concluding that EPA did not have the authority to issue it.[22] As a result, the rule is in effect while EPA conducts its review.

---

[19] For more information, see Executive Office of the President (EOP), *The President's Climate Action Plan*, June 2013; EOP, *Climate Action Plan: Strategy to Reduce Methane Emissions*, March 2014; EOP, "Fact Sheet: Administration Takes Steps Forward on Climate Action Plan by Announcing Actions to Cut Methane Emissions," January 14, 2015; and CRS Report R43860, *Methane: An Introduction to Emission Sources and Reduction Strategies*.

[20] GWP as calculated over 100 years. U.S. Environmental Protection Agency, *Inventory of U.S. Greenhouse Gas Emissions and Sinks: 1990-2013*, Washington, DC, EPA 430-R-15-004, April 15, 2015.

[21] In 2014, the United States had 6.87 billion metric tons of CO_2-equivalent emissions.

[22] *Clean Air Council v. Pruitt*, No. 17-1145, 2017 U.S. App. LEXIS 11803, at *28 (D.C. Cir. July 3, 2017).

AIR QUALITY STANDARDS

Background

Air quality has improved substantially since the passage of the Clean Air Act in 1970. Annual emissions of the six air pollutants for which EPA has set national ambient air quality standards (NAAQS)—ozone, particulate matter, sulfur dioxide, carbon monoxide, nitrogen dioxide, and lead—have declined by more than 70%, despite major increases in population, motor vehicle miles traveled, and economic activity.[23] Nevertheless, the goal of clean air continues to elude many areas, in part because scientific understanding of the health effects of air pollution has caused EPA to tighten standards for most of these pollutants. Congress anticipated that the understanding of air pollution's effects on public health and welfare would change with time, and it required, in Section 109(d) of the act, that EPA review the NAAQS at five-year intervals and revise them, as appropriate.

The most widespread air quality problems involve ozone and fine particles. A 2013 study by researchers at the Massachusetts Institute of Technology concluded that emissions of particulate matter and ozone caused 210,000 premature deaths in the United States in 2005.[24] Many other studies have found links between air pollution, illness, and premature mortality, as well. EPA summarizes these studies in what are called Integrated Science Assessments and Risk Analyses when it reviews a NAAQS, and, with input from the states, it identifies areas where concentrations of pollution exceed the NAAQS following its promulgation. As of December 31, 2017, 107 million people lived in areas classified as "nonattainment" for the ozone NAAQS; 23 million lived in areas that were nonattainment for the fine particle $_{(PM2.5)}$ NAAQS.[25]

Figure 1 identifies areas that had not attained one or more of the NAAQS as of December 2017.

2015 Revision of the Ozone NAAQS

Since 2008, review of the NAAQS for ozone has sparked recurrent controversy. A review completed in 2008 made the standards more stringent, but for the first time ever, the Administrator chose a health-based standard outside the range recommended by the independent scientific review committee established by the Clean Air Act to review the agency's work performed in the NAAQS-setting process. EPA suspended implementation

---

[23] For additional data on air pollution trends, see EPA's air trends website, at https://gispub.epa.gov/air/trendsreport/2016/.

[24] Fabio Caiazzo, et al., "Air Pollution and Early Deaths in the United States. Part I: Quantifying the Impact of Major Sectors in 2005," *Atmospheric Environment*, November 2013, pp. 198-208.

[25] Data are from the U.S. EPA "Green Book," at https://www3.epa.gov/airquality/greenbook/ popexp.html.

of the 2008 standard in September 2009 in order to consider further strengthening it, and proposed a more stringent standard in January 2010.[26]

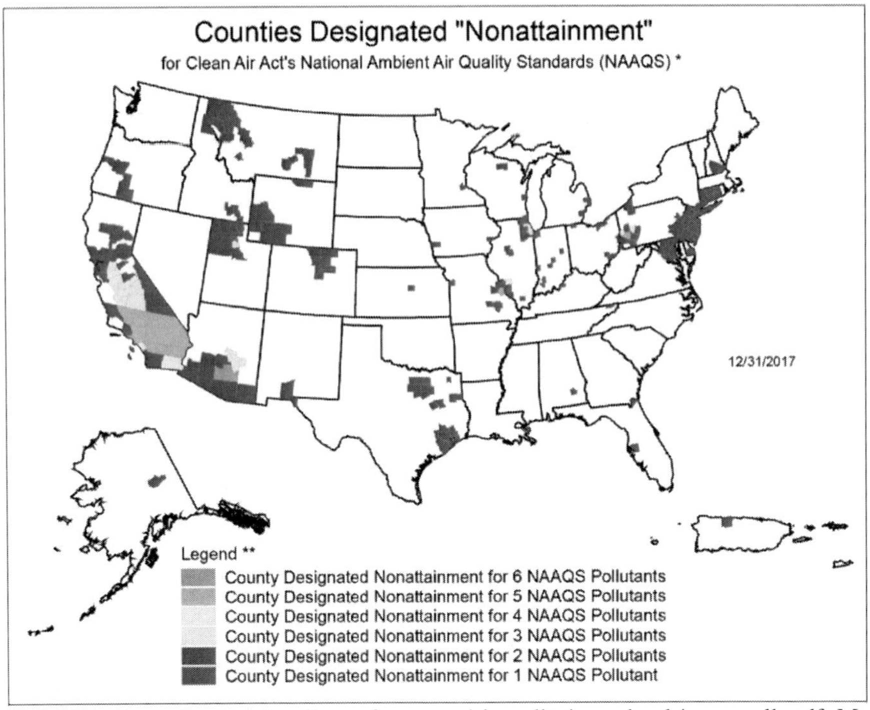

Source: U.S. EPA Green Book, https://www3.epa.gov/airquality/greenbook/mapnpoll. pdf. Map shows areas designated nonattainment as of December 31, 2017. Partial counties, those with part of the county designated nonattainment and part attainment, are shown as full counties on the map. In addition to the areas shown, two counties in Guam are designated nonattainment for the SO_2 NAAQS.

Figure 1. Counties Designated Nonattainment for One or More NAAQS.

On September 2, 2011, however, with a final rule in the last steps of interagency review at the Office of Management and Budget (OMB), the White House announced that President Obama had requested that the EPA Administrator withdraw the all-but-final (more stringent) ozone standards from further consideration at that time. The President's statement noted that "work is already underway to update a 2006 review of the science that will result in the reconsideration of the ozone standard in 2013," and stated that he did not "support asking state and local governments to begin implementing a new standard that will soon be reconsidered."[27] EPA then proceeded with the required five-year review of the 2008 standard, as the President indicated it would. The agency missed the statutory

---

[26] U.S. Environmental Protection Agency, "National Ambient Air Quality Standards for Ozone; Proposed Rule," 75 *Federal Register* 2938, January 19, 2010.

[27] The White House, Office of the Press Secretary, "Statement by the President on the Ozone National Ambient Air Quality Standards," September 2, 2011. For discussion of the 2010-2011 reconsideration, see CRS Report R42895, *Clean Air Issues in the 113th Congress: An Overview*.

deadline for completion of the review in March 2013, and a federal district court subsequently ordered the agency to release a final decision by October 1, 2015. The final standards were released on October 1, 2015, and appeared in the *Federal Register*, October 26, 2015.[28] The standards have been challenged in court; however, the court has paused the litigation while EPA conducts a review of the standards.[29]

The 2015 revision sets more stringent standards, lowering both the primary (health-based) and secondary (welfare-based) standards[30] from 75 parts per billion (ppb)—the level set in 2008—to 70 ppb. Using the latest available data, EPA identified 213 counties in 32 states outside California that had monitors showing nonattainment with the new standard in 2012-2014.[31] These are not the data EPA will use to designate nonattainment areas under the standard (EPA will use the most current monitoring data at the time, most likely data for 2014-2016 or later), but they served as the basis of EPA's analysis of the rule's potential effects. The agency's modeling shows all but 14 of these counties reaching attainment with a 70 ppb standard by 2025 as a result of already promulgated standards for power plants, motor vehicles, gasoline, and other emission sources.[32]

Thus, the agency's 2015 estimates of the cost of attaining a revised ozone NAAQS were substantially lower than many earlier estimates, including the agency's own estimate when it proposed a 70 ppb NAAQS in 2010. EPA estimated the cost of meeting a 70 ppb standard in all states except California at $1.4 billion annually in 2025. Because most areas in California would have until the 2030s to reach attainment,[33] EPA provided separate cost estimates for California ($0.8 billion in 2038). These cost estimates are substantially less than widely circulated estimates from the National Association of Manufacturers (NAM) and other industry sources.[34]

Members of Congress have shown particular interest in whether the expected benefits of the new ozone NAAQS will justify their costs. Both nationwide and in California, the agency said it expects the benefits of attainment to exceed the costs, but there is controversy over the methods used to estimate both. The agency prepares cost and benefit estimates at the time it proposes or promulgates a NAAQS—for information purposes and to comply with Executive Order 12866, under which the OMB requires cost-benefit analysis of economically significant rules.

---

[28] 80 *Federal Register* 65292. For links to the rule, as well as EPA's fact sheets and technical documents, see https://www.epa.gov/ozone-pollution/2015-national-ambient-air-quality-standards-naaqs-ozone.

[29] *Murray Energy Corp. v. EPA*, No. 15-1385 (D.C. Cir. Oct. 29, 2015).

[30] "Welfare" is defined by the statute to include effects on soils, water, crops, vegetation, man-made materials, weather, visibility, and climate, among other variables.

[31] See U.S. EPA, "County-Level Design Values for the 2015 Ozone Standards Based on Monitored Air Quality Data from 2012-2014," at https://www.epa.gov/sites/production/files/201603/documents/20151001datatable20122014.pdf. An additional 28 counties in California also have monitors showing nonattainment. EPA's analysis considered California separately, since most of the state's nonattainment areas will have until the late 2030s to reach attainment of the revised standard.

[32] See map at https://ozoneairqualitystandards.epa.gov/OAR_OAQPS/OzoneSliderApp/ index.html#.

[33] Under the statute, areas with more severe ozone pollution are given additional time to reach attainment of the standard, and must impose additional emission controls.

[34] For a discussion, see CRS Report R43092, *EPA's 2015 Ozone Air Quality Standards*.

As the Clean Air Act is currently written, however, the agency is prohibited from weighing costs against benefits in setting the standards. The Clean Air Act's Section 109 has been interpreted to prohibit consideration of costs in the setting of NAAQS since the provision was added to the act in 1970. In 2001, this interpretation was affirmed in a unanimous Supreme Court decision, *Whitman v. American Trucking Associations*.[35] Section 109 simply states that the EPA Administrator is to set the primary standard at a level requisite to protect public health, allowing an adequate margin of safety. The Court pointed to numerous other CAA sections where Congress had explicitly allowed consideration of economic factors, concluding that if Congress had intended to allow such factors in the setting of a primary NAAQS, it would have been more forthright—particularly given the centrality of the NAAQS concept to the CAA's regulatory scheme. The court concluded that Section 109(b)(1) "unambiguously bars cost considerations from the NAAQS-setting process."[36]

This is not to say that cost considerations play no role in Clean Air Act decisions, including in *implementation* of a NAAQS. Cost-effectiveness is considered extensively by EPA and the states in selecting emission control options to meet the standards. But in deciding what level of ambient pollution poses a health threat, the statute bars consideration of costs.

Congress has taken a keen interest in the results of the 2015 ozone review. Fifteen bills were introduced in the 114th Congress to modify EPA's authority or prohibit or delay the agency's proposed strengthening of the ozone NAAQS. Two of them, H.R. 4775 and H.R. 5538, passed the House, but neither bill was enacted. In the 115th Congress, H.R. 806 and S. 263 would, among other provisions, delay designation of nonattainment areas under the 2015 ozone standard until 2025 and submission of State Implementation Plans until 2026, require future NAAQS reviews every 10 years instead of every 5, and allow consideration of technological feasibility in the setting of some primary (health-based) standards. H.R. 806 is identical to one of the bills that passed the House in the 114th Congress, H.R. 4775. The House passed H.R. 806 on July 18, 2017. In addition, the House has passed H.R. 3354; Section 432 of the bill would delay implementation of the 2015 ozone NAAQS until the mid-2020s.

On June 6, 2017, EPA Administrator Pruitt announced a one-year delay in designating nonattainment areas under the revised NAAQS. On August 2, however, he reversed himself, stating that EPA "now understands that the information gaps that formed the basis of the extension may not be as expansive as we previously believed." The wording he chose continued to leave open the possibility that the October 2017 deadline would be extended,

---

[35] 531 U.S. 457 (2001).

[36] For further discussion of the American Trucking case, see CRS Report RS20860, *The Supreme Court Upholds EPA Standard- Setting Under the Clean Air Act: Whitman v. American Trucking Ass'ns*, by Robert Meltz and James E. McCarthy.

at least for some areas; and, in fact, as of mid-March 2018, nonattainment areas have not been designated for the 2015 standard.

OTHER ISSUES

At the outset of the 115th Congress, with a Republican President and Republican majorities in both the House and Senate, much was made of the opportunity presented to override EPA regulatory actions from the final days of the Obama Administration through use of the Congressional Review Act (CRA). The CRA provides expedited procedures that can be used during a limited period of time after promulgation of an agency rule to bring resolutions disapproving regulations to the Senate floor, where they cannot be amended or filibustered. In doing so, the CRA can remove commonly used obstacles to Senate action.

A December 15, 2016, CRS analysis concluded that joint resolutions disapproving regulations promulgated and received in Congress on or after June 13, 2016, were eligible to use these fast-track Senate procedures during the first 60 legislative days of the 115th Congress.[37] A list of major rules promulgated on or after June 13 included six Clean Air Act rules promulgated by EPA: the Renewable Fuel Standard (RFS) Program Standards for 2017 and Biomass-Based Diesel Volume for 2018; Formaldehyde Emission Standards for Composite Wood Products; the Cross-State Air Pollution (CSAPR) Update Rule; the Phase 2 Medium- and Heavy-Duty Truck GHG Emission Rule; New Source Performance Standards for Methane Emissions from Municipal Solid Waste (MSW) Landfills; and Emission Guidelines for Existing MSW Landfills.[38] The CRA's fast track authority expired on May 11, 2017, without any of these rules being disapproved.

These and other rules—those promulgated and received in Congress before June 13, 2016—could still be addressed by Congress through targeted legislation or through the appropriations process. In the 115th Congress, the House Energy and Commerce Committee reported three bills that would overturn or delay specific EPA air regulations: H.R. 1917, the BRICK Act, which would delay implementation of hazardous air pollutant emission standards affecting brick and ceramic manufacturers until all legal challenges to the rules are settled; H.R. 1119, the SENSE Act, which would ease emission limits for electric generating units powered by coal refuse; and H.R. 453, which would delay the implementation of Phase 2 emission standards for wood heaters to 2023. The House passed H.R. 1119 and a bill that combined the other two on March 8, 2018.

These and other bills would address criticism by industry groups and others that EPA under previous Administrations has overreached its authority. Environmental and public

---

[37] CRS Insight IN10437, *Agency Final Rules Submitted on or After June 13, 2016, May Be Subject to Disapproval by the 115th Congress.*

[38] See Maeve P. Carey et al., "Major" Obama Administration Rules Potentially Eligible to be Overturned under the Congressional Review Act in the 115th Congress," CRS General Distribution Memorandum, January 31, 2017.

health groups, on the other hand, generally believe that the agency has not overreached in setting Clean Air Act standards. These groups often maintain that the agency's standards are not stringent enough, do not meet statutory requirements, or disregard the findings of the agency's science advisors. The result is that EPA Clean Air Act standards generally are challenged in court both by industry and by environmental groups, with various states supporting each side. The resulting court decisions often set EPA's agenda as much as Congress or the Administration.

The courts will continue to play an important role, and the 115th Congress and EPA itself may act to make important revisions to Clean Air Act regulations. In short, the CAA's regulatory structure continues in a state of uncertainty in 2018, with possible changes emanating from the executive, legislative, and judicial branches.[39]

---

[39] For additional discussion of EPA's regulatory actions under all the environmental statutes, see CRS Report R41561, *EPA Regulations: Too Much, Too Little, or On Track?*

In: The Clean Air Act and The Clean Power Plan
Editor: Claire Ferguson
ISBN: 978-1-53614-825-1

Chapter 2

COST AND BENEFIT CONSIDERATIONS IN CLEAN AIR ACT REGULATIONS*

SUMMARY

The Clean Air Act (CAA) gives the Environmental Protection Agency (EPA) broad authority to set ambient air quality standards to protect public health and welfare. It authorizes emission standards for both mobile and stationary air pollution sources, including cars, trucks, factories, power plants, fuels, consumer products, and dozens of other source categories. Since 1970, EPA has used this authority to require emission controls for these sources. Emissions of the most widespread ("criteria") pollutants have been reduced by 72% during that period.

As directed by Congress and by executive orders, EPA has estimated the costs and benefits of major CAA (and other) regulations for the last four decades. Its most comprehensive recent studies and studies by the Office of Management and Budget (OMB) have concluded that the benefits of clean air regulations outweigh the costs by substantial margins. EPA's cost-benefit analyses of individual regulations, required by Executive Order 12866, show similar results: a review of the 55 economically significant CAA regulations promulgated from 2001 to 2016 found only two in which estimated costs exceeded benefits.

Nevertheless, many in Congress have expressed concern that Clean Air Act and other environmental regulations harm the nation's economy. One issue raised by critics is whether EPA underestimates the cost and other negative impacts of CAA rules—in part, by considering them individually, and not considering cumulative impacts. Another criticism is that the agency relies for most of its benefit assessments on the effects of reducing a single category of pollutants, particulate matter (PM). Research has tied PM to tens of thousands of premature deaths, and EPA often finds that reductions in PM emissions justify regulation, even where PM reductions are a "co-benefit" of reducing another targeted pollutant. A third issue critics raise is whether the methodology used to place

* This is an edited, reformatted and augmented version of a Congressional Research Service report, Publication No. R44840, dated May 5, 2017.

monetary value on the avoidance of premature death—a technique referred to as calculating the "value of a statistical life"—inflates the estimated benefits of regulation.

This chapter examines these issues in the context of Clean Air Act regulation. It reviews EPA and Office of Management and Budget (OMB) studies of the cost and benefit of CAA regulations, and addresses the issues raised by agency critics. The report finds that

- The Clean Air Act authorizes EPA to set standards in multiple sections of the act: about half of the act's major regulatory authorities mention costs or economic considerations explicitly, and several others imply that costs may be considered; but other authorizing sections, including some key sections, make no mention of cost considerations.
- Where the statutory authorities do not mention cost consideration, they tend to fall into one of four categories: provisions in which Congress itself set the standards; provisions where Congress directed the agency to set health-based standards, without mentioning cost; broad authority to promulgate regulations to achieve an objective that Congress determined was necessary, but the specifics of which it could not anticipate; or authority to promulgate federal requirements in cases where states have failed to develop or implement adequate regulations on their own to meet a federal mandate.
- In all cases, even where the statute would prohibit consideration of cost in setting standards, EPA is bound by executive orders to provide estimates of costs and benefits if the rule would be economically significant.
- According to EPA, the estimated benefits of CAA regulation will exceed the estimated costs by more than 30 to 1 in the period 1990-2020. CAA regulations prevent 230,000 premature deaths annually, according to the agency.
- The estimated benefits of CAA regulations rely heavily on the effects of reducing particulate emissions, and on the value placed on the avoidance of premature death as a result of such controls.
- Many rules have benefits or costs that cannot be quantified or monetized in light of existing information.

President Trump has issued two executive orders that address the cost of EPA regulations: Executive Order (E.O.) 13771, signed January 30, 2017, and E.O. 13783, signed March 28, 2017. The former directs OMB to set regulatory "budgets" for executive branch departments and agencies and, in general, to rescind two regulations for every new one issued. The latter requires EPA to review—and, if appropriate, suspend, revise, or rescind—several CAA regulations affecting energy production, with an eye to avoiding regulatory burdens. At present, the effect of the two orders on future CAA regulations is unclear. The report discusses some of the possible implications.

INTRODUCTION

In the past several Congresses, critics of the Environmental Protection Agency (EPA) have focused much of their attention on the agency's regulatory actions under the Clean Air Act (CAA). During this time, the Obama Administration promulgated numerous CAA regulations. In general, these regulations came in response to congressional authority or mandates under the Clean Air Act Amendments of 1970, 1977, and 1990. A number were also in response to court decisions that remanded to the agency regulations that the agency had promulgated—often under previous administrations. Critics have maintained that many of these regulatory actions would be too costly, harming a wide range of industries, and would not be justified by the benefits obtained.[1]

The Clean Air Act is not consistent in whether it allows or requires the consideration of costs and benefits in setting standards. (See "Clean Air Act Authorities," below.) The act requires or authorizes the EPA Administrator to promulgate regulations or set standards in more than 60 sections or subsections. In 25 of these sections or subsections, cost is not mentioned or implied as a factor to be considered. In the remaining sections and subsections where the Administrator is required or authorized to promulgate regulations—more than 40 of them in all—cost *is* either identified explicitly or implied as a factor to be considered.

Whether or not the statute requires a consideration of cost, EPA has prepared cost estimates for all economically significant rules since the Carter Administration as the result of executive orders.[2] Under Executive Order (E.O.) 12866, each economically significant regulatory action taken by Executive Branch agencies (under any statutory authority) must include estimates of the cost and benefits of the action in a Regulatory Impact Analysis (RIA) before it is proposed, and again before it is promulgated.[3] RIAs play a major role in the interagency review process overseen by the Office of Management and Budget, which precedes the publication of significant rules in the *Federal Register*.

Thus, there is a process for considering the costs and benefits of all Clean Air Act (and other) economically significant regulations. How well this process works is the question. One issue raised by EPA's critics is whether the agency underestimates the cost and other negative impacts of rules in these RIAs by considering them individually, and not considering cumulative impacts. A second criticism is that the agency relies for most of its CAA benefit assessments on the effects of reducing a single category of pollutants, particulate matter (PM). Research has tied PM to tens of thousands of premature deaths, and EPA often finds that reductions in PM emissions justify regulation, even where the target of the regulations is a different pollutant. A third issue critics raise is whether the methodology used to place monetary value on the avoidance of premature death—a technique referred to as calculating the "value of a statistical life"—inflates the estimated benefits of regulation. A fourth issue, not generally raised by critics, but often noted by

[1] See, for example, U.S. Chamber of Commerce, "Regulatory Areas: Energy and Environment," at https://www.uschamber.com/regulations/areas. Of the 38 EPA regulations listed by the Chamber, 24 are Clean Air Act rules.

[2] The Carter Administration order, "Improving Government Regulations," Executive Order 12044, signed March 23, 1978, is at http://www.thecre.com/pdf/12044.PDF.

[3] "Regulatory Planning and Review," Executive Order 12866, signed September 30, 1993, at http://www.plainlanguage.gov/populartopics/regulations/eo12866.pdf. For a discussion of E.O. 12866 and the regulatory review process, see CRS Report RL32397, *Federal Rulemaking: The Role of the Office of Information and Regulatory Affairs*, coordinated by (name redacted).

regulators, concerns the difficulty of identifying and estimating the full range of costs and benefits of environmental regulations.

This chapter examines these issues in the context of Clean Air Act (CAA) regulations. It begins with a review of EPA's CAA authorities and the role of cost considerations in CAA standard-setting.

Clean Air Act Authorities

The Clean Air Act, originally enacted in 1955 and amended numerous times since then, gives EPA sweeping powers to "protect and enhance the quality of the Nation's air resources so as to promote the public health and welfare and the productive capacity of its population." In the statute, Congress directs the EPA Administrator to, among other things

- set national ambient air quality standards;
- set emission standards for both stationary and mobile sources of air pollution;
- reduce emissions of 187 hazardous air pollutants that Congress itself listed in the statute;
- protect air quality in relatively pristine areas from significant deterioration;
- regulate fuels and fuel additives, both to protect public health and welfare and to prevent the impairment of emission control devices;
- require the use of renewable transportation fuels;
- control acid deposition;
- protect the stratospheric ozone layer by requiring the phase-out of ozone-depleting substances;
- issue permits and enforce the act's emission limits; and
- develop and enforce Federal Implementation Plans in states that fail to implement the act's requirements.

The specific authorities given to the Administrator are established in more than 60 different sections and subsections of the act, which range from broad authority to protect public health with an adequate margin of safety to detailed requirements that specify numerical emission limits or require that standards be at least as stringent as the emission limitation achieved by the best controlled similar source.[4]

---

[4] A list of the regulatory authorities that the act grants the EPA Administrator and the relevant language of each is available from the authors.

Cost Consideration in Clean Air Act Authorities

Because the act's authorities are so fragmented, it can be difficult to generalize regarding the role of cost considerations in setting air quality standards. Many of the act's authorities allow or require the Administrator to "take into account" or "take into consideration" the cost or technical feasibility of specific emission requirements.

Authorities That Mention or Imply Cost Considerations

A review of 67 sections, subsections, or provisions of the act that authorize regulations indicates that about half (34) specifically mention cost or economic considerations. Among the 34 sections are several major regulatory authorities, including the authority to

- set emission standards for new stationary sources (power plants, refineries, etc.) in Section 111;
- go "beyond the floor" in emission standards for sources of 187 hazardous air pollutants, under Section 112(d);
- set emission standards for motor vehicles (beyond the standards specifically listed in the act), under Sections 202(a) and 202(i);
- control mobile source air toxics, under Section 202(l);
- control or prohibit the manufacture and sale of fuels and fuel additives under Section 211(c);
- require the sale of reformulated gasoline in nonattainment areas, under Section 211(k);
- set emission standards for nonroad vehicles and engines, including construction equipment, recreational equipment, agricultural machinery, electric generators, and other sources, under Section 213; and
- set emission standards for locomotives, buses, and aircraft, under Sections 213, 219, and 231.

In eight other cases, consideration of cost is implied by the act, e.g., where it requires a standard that is "practicable" or "reasonably achievable." These sections of the act direct the EPA Administrator to

- consider the "remaining useful life of the existing source" to which an emission standard will apply, under Section 111(d);
- provide for the use of "generally available control technologies" to control area sources of hazardous air pollutants, under Section 112(d)(5);

- promulgate "reasonable regulations and appropriate guidance to provide, to the greatest extent practicable, for the prevention and detection of accidental releases" of extremely hazardous substances and take into consideration "the concerns of small business," under Section 112(r)(7);
- consider "the availability and feasibility of pollution control measures" in classifying nonattainment areas under Section 172;
- consider "such other factors as he [the Administrator] deems pertinent" and take into consideration "the restraints of an adequate lead time for design and production" in setting vapor recovery standards for gasoline under Section 202(a)(5)
- impose emissions standards or emissions control technology requirements that "reflect the best retrofit technology and maintenance practices reasonably achievable" for retrofit of urban buses under Section 219(d);
- decide whether a requirement is "practicable, taking into account technological achievability, safety, and other relevant factors" in establishing an accelerated schedule for phasing out production and consumption of ozone-depleting substances under Section 606; and
- consider "the purpose or intended use of the product, the technological availability of substitutes ..., safety, health, and other relevant factors" in regulating nonessential products that release class I ozone depleting substances[5] under Section 610 (except for two specific categories of products that are listed in the statute).

A full list of the 42 provisions that mention or imply consideration of cost is provided in Table 1.

Table 1. Clean Air Act Authorizing Provisions That Mention or Imply Consideration of Cost

Section	Provision
111 (a) and (b)	New Source Performance Standards (NSPS)
111(d)	Existing Source Performance Standards
111(h)	Work Practice Standards in Lieu of NSPS or Section 111(d) Standards
112(d)(2) and (3)	Maximum Achievable Control Technology Standards for New Sources of Hazardous Air Pollutants[a]
112(d)(2) and (3)	Maximum Achievable Control Technology Standards for Existing Sources of Hazardous Air Pollutants[a]
112(d)(5)	National Emission Standards (NESHAP) for Area Sources of Hazardous Air Pollutants
112(d)(8)	National Emission Standards (NESHAP) for Coke Ovens[a]
112(f)	Residual Risk Standards to prevent adverse environmental effects

[5] Class I substances, 20 of which are listed in Section 602(a) of the act, are defined by their potential to damage the Earth's ozone layer.

Section	Provision
112(h)	NESHAP Work Practice Standards
112(k)	Area Source Program for Hazardous Air Pollutants
112(m)	Great Lakes and Coastal Waters
112(r)(7)	Accident Prevention
119(b)	Primary Nonferrous Smelter Orders
125	Measures to Prevent Economic Disruption or Unemployment
129(a)(2)	Solid Waste Combustion[a]
169A	Visibility Protection
172	Classification of Nonattainment Areas
183(e)	Control of Emissions from Consumer or Commercial Products
183(f)(1)	Standards for Loading or Unloading Tank Vessels
202(a)	Emission Standards for New Motor Vehicles
202(a)(3)(A) and (B)	Emission Standards for Heavy Duty Trucks
202(a)(3)(D)	Emission Standards for Rebuilt Engines
202(a)(5)	Gasoline Vapor Recovery
202(f)(3)	High Altitude Emission Standards
202(i)	Tier 2 Auto Emission Standards
202(k)	Evaporative Emission
202(l)(2)	Mobile Source Air Toxics
211(c)	Regulation of Fuels or Fuel Additives that Endanger Public Health or Welfare
211(c)	Regulation of Fuels or Fuel Additives that Impair the Performance of Emission Controls
211(h)	Reid Vapor Pressure Requirements
211(k)	Reformulated Gasoline for Conventional Vehicles
211(k)	Toxic Air Pollutant Emissions from Reformulated Gasoline
213(a)	Emission Standards for Nonroad Engines and Vehicles
213(a)	Emission Standards for Locomotives
219(a)	Urban Bus Standards
219(d)	Urban Bus Retrofit Requirements
231(b)	Aircraft Emission Standards
245	Standards for Heavy-Duty Clean-Fuel Vehicles
246	Centrally Fueled Fleets
407(b)(2)	Nitrogen Oxide Emission Reductions
606	Accelerated Phase-Out Schedule for Class I and Class II Substances
610	Nonessential Products Containing Chlorofluorocarbons(for products identified by the Administrator)[b]

Source: CRS analysis of the Clean Air Act.

[a] Section 112(d)(2) and Section 129(a)(2) both modify the Administrator's authority to determine what is "achievable" by stating "taking into consideration the cost of achieving such emission reduction, and any non-air quality health and environmental impacts and energy requirements." In both cases, a statement that the standard "shall not be less stringent than..." follows the mention of cost, apparently requiring the Administrator to promulgate standards not less stringent than the best similar source (in the case of new sources) or the average of the top 12% (in the case of existing sources), regardless of cost. The statement regarding cost considerations would appear to apply only in cases where the Administrator promulgates a standard that goes beyond the minimum ("not less stringent than") floor. Similarly, Section 112(d)(8) modifies the Administrator's authority to establish coke oven standards by directing him to take into account costs and reasonable commercial door warranties, but the statement that, "Such regulations shall require at a minimum [followed by a list of specific standards]" follows the mention of cost, apparently requiring the Administrator to promulgate standards that meet the minimum listed requirements regardless of cost.

[b] Section 610 states that the Administrator shall promulgate regulations that prohibit any person from selling or distributing two specific types of chlorofluorocarbon (CFC)-containing products. There is no mention of cost in connection with the regulation of these two types of product. The section also directs the Administrator to identify other nonessential CFC-containing products for such a prohibition. In the latter case, the Administrator is to consider the "technological availability of substitutes for such product and for such class I substance, safety, health, and other relevant factors," implying that cost may be considered. Thus, this authority, like that identified in note a., allows or requires the consideration of cost in some instances, but not in others, and the provision is listed in both Table 1 and Table 2 of this chapter.

Authorities That neither Mention nor Imply Cost Consideration

In 25 CAA sections or subsections where regulatory authority is conferred on the Administrator (identified in Table 2), cost is not mentioned or implied as a factor to be considered. These statutory authorities tend to fall into one of four categories: (1) provisions in which Congress itself set the standards; (2) provisions where Congress directed the agency to set health-based standards, without mentioning cost; (3) provisions in which Congress gave the agency broad authority to promulgate regulations to achieve an objective that Congress determined was necessary (generally protecting public health directly or indirectly, or protecting the environment), but the specifics of which Congress could not anticipate; or (4) a provision requiring EPA to promulgate federal requirements in cases where states have failed to develop or implement adequate regulations to meet a federal mandate. These authorities are discussed briefly in the sections that follow.

Standards Established by Congress

In the late 1980s, when the most recent major CAA amendments[6] were being drafted, Congress, frustrated with the slow progress being achieved under earlier versions of the statute and by the delays caused by litigation, limited EPA's discretion in setting emission standards in a number of cases by writing detailed regulatory requirements into the statute.[7]

Perhaps the most specific of these requirements appear in Section 202 of the act, which addresses motor vehicle emissions. Here, Congress listed the pollutants to be controlled, mandated specific numerical standards for their emission, and set schedules for implementation. EPA still needed to promulgate regulations to implement these standards, but the standards themselves were set by Congress to take effect on a date certain.

Another major example can be found in Section 112, where Congress addressed emissions of hazardous air pollutants. Here, Congress listed 187 pollutants the emissions of which were to be controlled; defined the threshold quantity of emissions that would require sources to meet the most stringent standards; and required that sources meet emission limits at least as stringent as the emissions of the best controlled similar sources. The act mandated a 10-year schedule for promulgating standards.

Health-Based Standards

The cornerstone of the Clean Air Act consists of health-based standards for widespread air pollutants identified by EPA under Sections 108 and 109 of the act. These standards, termed National Ambient Air Quality Standards (NAAQS) are for air pollutants that, in the Administrator's judgment, "endanger public health or welfare" and "the presence of which

[6] The Clean Air Act Amendments of 1990, P.L. 101-549.

[7] Congress included detailed prescriptive standards and deadlines in other environmental legislation in the 1980s, notably the Hazardous and Solid Waste Amendments of 1984 (P.L. 98-616).

in the ambient air results from numerous or diverse mobile or stationary sources."[8] The primary (health-based) NAAQS must be designed to protect public health with an adequate margin of safety.

Table 2. Clean Air Act Authorizing Provisions that Do Not Mention Cost

Section	Provision
109(a)	Primary National Ambient Air Quality Standards
109(b)	Secondary National Ambient Air Quality Standards
110(c)(1)	Federal Implementation Plans
112(d)(2) and (3)	Maximum Achievable Control Technology Standards for New Sources of Hazardous Air Pollutants[a]
112(d)(2) and (3)	Maximum Achievable Control Technology Standards for Existing Sources of Hazardous Air Pollutants[a]
112(d)(8)	Coke Oven Standards[a]
112(f)	Hazardous Air Pollutant Residual Risk Standardsto provide an ample marginof safety to protect public health
112 (k)	Area Source Program for Urban Hazardous Air Pollutants
126(c)	Interstate Pollution Abatement
129(a)(2)	Solid Waste Combustion[a]
166	Prevention of Significant Deterioration
202(a)	Onboard Vapor Recovery (for motor vehicles)
202(j)	Cold CO [Carbon Monoxide]Standard
202(m)	EmissionsControl Diagnostics
211(i)	Sulfur Content of Diesel Fuel
211(o)	Renewable Fuels
219(c)	Urban Bus Low-Polluting Fuel Requirement
243	Emission Standards for Light-Duty Clean Fuel Vehicles
328	Air Pollution from Outer Continental Shelf Activities
407(b)(1)	Nitrogen Oxides Emission Reduction Program
604(c)	Production and Consumption of Class I (Ozone-Depleting) Substances
605(c)	Production and Consumption of Class II (Ozone-Depleting) Substances
609	Servicing of Motor Vehicle Air Conditioners
610	Nonessential Products Containing CFCs (for listed categories)
615	Authority of Administrator(regarding protection of the stratosphere)

Source: CRS analysis of the Clean Air Act.

[a] As noted above in footnote a. to Table 1, Section 112(d)(2), Section 112(d)(8), and Section 129(a)(2), all modify the Administrator's authority to set standards by requiring that he take costs into account, but also set minimum requirements for the standards that must be met regardless of cost. As a result, these authorities are listed in both Table 1 and Table 2 of this chapter.

[b] As noted above in footnote b. to Table 1, Section 610 requires the Administrator to promulgate regulations that prohibit any person from selling or distributing two specific types of chlorofluorocarbon (CFC)-containing products. There is no mention of cost in connection with the regulation of these two types of product. The section also directs the Administrator to identify other nonessential CFC-containing products for such a prohibition. In the latter case, the Administrator is to consider the "technological availability of substitutes for such product and for such class I substance, safety, health, and other relevant factors," implying that cost may be considered. As a result, the Section 610 authorities are listed in both Table 1 and Table 2 of this chapter.

[8] Section 108(a)(1)(A) and (B).

Using this authority, EPA has promulgated NAAQS for six air pollutants or groups of pollutants: sulfur dioxide (SO_2), particulate matter ($PM_{2.5}$ and PM10), nitrogen dioxide (NO2), carbon monoxide (CO), ozone, and lead. The act requires EPA to review the scientific data upon which the standards are based, and revise the standards, if necessary, every five years. NAAQS do not directly regulate emissions or directly compel actions by sources of pollution. In essence, they are standards that define what EPA considers to be clean air for the specified pollutants. Once a NAAQS has been set, the agency, using monitoring data and other information submitted by the states, identifies areas that exceed the standard and that must, therefore, reduce pollutant concentrations to achieve it. After these "nonattainment" areas are identified, state and local governments have up to three years to produce State Implementation Plans that outline the measures they will implement to reduce the pollution levels and attain the standards.

The issue of cost is a perennial one in NAAQS decisions. For 45 years, EPA has interpreted Section 109 as prohibiting the Administrator from considering costs in setting the standards. In 2001, this interpretation was affirmed in a unanimous Supreme Court decision, *Whitman v. American Trucking Associations*.[9] The Court pointed to numerous other CAA sections where Congress had explicitly allowed consideration of economic factors, concluding that if Congress had intended to allow such factors in the setting of a primary NAAQS, it would have been more forthright—particularly given the centrality of the NAAQS concept to the CAA's regulatory scheme. The court concluded that Section 109(b)(1) "unambiguously bars cost considerations from the NAAQS-setting process."[10]

Broad Authority

A third group of Clean Air Act standards are the result of Congress giving EPA authority to promulgate regulations to achieve an objective (generally, protecting public health directly or indirectly, or protecting the environment) that Congress determined was essential, but the specifics of which it might not have been able to anticipate. This authority is similar to that for health-based standards, but broader: it can be used to protect the environment or what the act defines as "welfare,"[11] in addition to public health, and it authorizes controls of specific substances and activities.

For example, under Title VI of the Clean Air Act, Congress directed the EPA Administrator to phase out the production and consumption of chemicals identified as Class I Ozone Depleting Substances (ODS). ODS affect the stratospheric ozone layer,

---

[9] 531 U.S. 457 (2001).

[10] For further discussion of the American Trucking case, see CRS Report RS20860, *The Supreme Court Upholds EPA Standard- Setting under the Clean Air Act: Whitman v. American Trucking Ass'ns*, by (name redacted) and (name red acted).

[11] The act defines welfare in Section 302(h): "All language referring to effects on welfare includes, but is not limited to, effects on soils, water, crops, vegetation, man-made materials, animals, wildlife, weather, visibility, and climate, damage to and deterioration of property, and hazards to transportation, as well as effects on economic values and on personal comfort and well-being, whether caused by transformation, conversion, or combination with other air pollutants."

which protects the Earth from harmful radiation. A list of ODS was provided in the statute. In addition, EPA was directed in Section 602(a) to add to the list "any other substance that the Administrator finds causes or contributes significantly to harmful effects on the stratospheric ozone layer." A schedule for the phase-out of these chemicals is provided in Section 604(c). There is no discussion of economic impact or cost in these sections.

Section 615 of the act is even broader:

> If, in the Administrator's judgment, any substance, practice, process, or activity may reasonably be anticipated to affect the stratosphere, especially ozone in the stratosphere, and such effect may reasonably be anticipated to endanger public health or welfare, the Administrator shall promptly promulgate regulations respecting the control of such substance, practice, process, or activity, and shall submit notice of the proposal and promulgation of such regulation to the Congress.

There is no mention of cost or economic impact in the section.

Federal Implementation Plans

The fourth group of regulations for which the Clean Air Act does not require consideration of cost is the result of backup authority that Congress gave EPA. In general, the act envisions that states will be responsible for adopting regulations to attain National Ambient Air Quality Standards. Section 110 of the act discusses in great detail the implementation plans that states are to submit to EPA, describing how they will attain or maintain compliance with the NAAQS. EPA cannot compel a state to submit a State Implementation Plan. Rather, if a state fails to submit a satisfactory plan by the statutory deadline, or fails to correct a deficiency identified by the EPA Administrator, EPA is required to promulgate a Federal Implementation Plan for the state under Section 110(c) of the act. The statute provides that

> The Administrator shall promulgate a Federal implementation plan at any time within 2 years after the Administrator—
>
> (A) finds that a State has failed to make a required submission or finds that the plan or plan revision submitted by the State does not satisfy the minimum criteria established under subsection (k)(1)(A) of this section, or
>
> (B) disapproves a State implementation plan submission in whole or in part, unless the State corrects the deficiency, and the Administrator approves the plan or plan revision, before the Administrator promulgates such Federal implementation plan.

There is no directive for the Administrator to consider cost or economic impact in developing such a plan.

EPA's Use of Cost-Benefit Analysis

Although the statute prohibits the consideration of cost in setting some standards, EPA is subject to executive orders that require the estimation of costs and benefits any time an agency develops "economically significant" regulations. Executive Order 12866 defines an "economically significant" regulation as any rule that may "have an annual effect on the economy of $100 million or more or adversely affect in a material way the economy, a sector of the economy, productivity, competition, jobs, the environment, public health or safety, or State, local, or tribal governments or communities."[12] The term "effect on the economy" means that a rule may be considered economically significant if it has costs *or* benefits of over $100 million.[13] The cost and benefit estimates are to be provided by the regulatory agency before rules are proposed for public comment, and again before they are issued in final form. E.O. 12866 states that, "Each agency shall assess both the costs and the benefits of the intended regulation and, recognizing that some costs and benefits are difficult to quantify, propose or adopt a regulation only upon a reasoned determination that the benefits of the intended regulation justify its costs."

OMB has issued a number of guidance documents that agencies are required to follow when estimating costs and benefits of regulations. OMB's most significant guidance document is Circular A-4 on "Regulatory Analysis."[14] The circular states that it was "designed to assist analysts in the regulatory agencies by defining good regulatory analysis ... and standardizing the way benefits and costs of Federal regulatory actions are measured and reported." EPA's agency-specific document, "Guidelines for Preparing Economic Analyses (2010),"[15] is built on the analytical framework of Circular A-4.

Circular A-4 states that a "good regulatory analysis should include the following three basic elements: (1) a statement of the need for the proposed action, (2) an examination of alternative approaches, and (3) an evaluation of the benefits and costs—quantitative and qualitative—of the proposed action and the main alternatives identified by the analysis."[16]

With regard to analytical approaches, the circular states that agencies should use both cost-benefit analysis (CBA) and cost-effectiveness analysis. Cost-benefit analysis, in this context, involves the systematic identification of all of the costs and benefits associated with a forthcoming regulation, including nonquantitative and indirect costs and benefits, and how those costs and benefits are distributed across different groups in society. Cost-effectiveness analysis seeks to determine how a given goal can be achieved at the least

---

[12] "Regulatory Planning and Review," Executive Order 12866, 58 *Federal Register* 51735, October 4, 1993.

[13] The phrase also includes transfer rules that transfer sums of over $100 million.

[14] OMB Circular A-4, "Regulatory Analysis," September 17, 2003, at http://www.whitehouse.gov/omb/assets/regulatory_matters_pdf/a-4.pdf. The circular took effect for "economically significant" proposed rules on January 1, 2004, and for "economically significant" final rules on January 1, 2005.

[15] See EPA, *Guidelines for Preparing Economic Analyses*, December 17, 2010, at https://www.epa.gov/environmentaleconomics/guidelines-preparing-economic-analyses.

[16] OMB Circular A-4, p. 2.

cost. When all benefits and costs can be expressed in monetary units, CBA can clearly indicate which approach is most efficient in terms of net benefits. However, in many (and perhaps most) cases, agencies are not able to express all of the benefits or costs in monetary units. In such cases, cost-effectiveness analysis is available to consider the most economically efficient approaches.

Measuring Costs and Benefits

In ideal circumstances, regulations should be designed to maximize "net benefits" (that is, maximizing the value of "total benefits" minus "total costs"). To better assess net benefits, it is important to understand the market failures and economic externalities for which environmental regulations (including air emissions regulations) try to correct.

Economic Externalities, Market Failures, and Government Regulation

Market failure describes an economic situation where the individual incentives for rational behavior do not lead to rational outcomes for the society. Market failure may arise from various economic circumstances, including externalities, monopolies, public goods, and information asymmetries.

Environmental problems are a classic case of economic externality. Externalities arise when an individual, a firm, or a government takes an action but does not bear all the costs (negative externality) or all the benefits (positive externality) of the action. An example of a negative externality would be a factory that pollutes as a result of its production process. This pollution may pose health risks for nearby residents or degrade the quality of the air or water. However, the additional cost to address any health issues or to help maintain the cleanliness of the air or water are not directly paid by the owner of the factory.

Many economists argue that the existence of externalities justifies government intervention in the markets through legislation or regulation. Ideally, regulation would be designed to get individuals or firms to "internalize the externality" by considering the full costs of production. In outlining this consideration, E.O. 12866 states that "Federal agencies should promulgate only such regulations as are required by law, are necessary to interpret the law, or are made necessary by compelling need, such as material failures of private markets to protect or improve the health and safety of the public, the environment, or the wellbeing of the American people." Many environmental regulations are designed to require firms to "internalize the externality" through a variety of policy instruments (e.g., performance standards, taxes and fees, or trading schemes).

Analyzing environmental regulations in terms of market theory can help show that the consideration of costs and benefits are often the opposite side of the same coin (See Text Box). OMB's circular and EPA's guidelines summarize a variety of methods the agency can use to determine the total costs and total benefits of a regulation—including those which are difficult to quantify and monetize.

- In measuring costs, the guidance documents ask the agency to assess the direct costs to the regulated firms, including pollution control equipment, record keeping and reporting requirements and labor for equipment installation, operation, maintenance, and monitoring. Further, the agency should attempt to analyze additional and/or indirect impacts on consumers, small businesses, government entities (including administrative cost and savings), international trade, and energy and employment effects.[17]
- In measuring benefits, the guidance documents ask the agency to use the best reasonably obtainable scientific, technical, economic, and other information available to quantify—and, if possible, monetize—the impacts of regulations. For example, the benefits of a regulation that reduces emissions of air pollution might be quantified in terms of a variety of health, climate, visibility, and ecosystem effects. Such benefits may include the number of premature deaths avoided each year; the number of prevented nonfatal illnesses and hospitalizations; the number of prevented lost work or school days; improvements in visibility in specific regions; and improvements in ecosystem health as measured by specific indicators (e.g., lake acidification). These quantified benefits may be monetized using a number of tools and indicators, including EPA's Environmental Benefits Mapping and Analysis Program and various metrics such as Value of Statistical Life, Quality-Adjusted Life Years, and Social Cost of Carbon,[18] among others.
- Some benefits are difficult to quantify and monetize. In such cases, the guidance documents ask the agency to include a qualitative discussion of benefits results. The discussion should explain why a quantitative analysis was not possible and the reasons for believing that these non-quantified effects may be important for decision making.

---

[17] EPA uses a variety of methods, models, and tools to estimate economic impacts. See "Economic and Cost Analysis for Air Pollution Regulations," at https://www.epa.gov/economic-and-cost-analysis-air-pollution-regulations.

[18] On March 28, 2017, President Trump issued Executive Order 13783, "Promoting Energy Independence and Economic Growth." It states that "it is essential that agencies use estimates of costs and benefits in their regulatory analyses that are based on the best available science and economics." His order then effectively withdrew the federal "social costs of greenhouse gases" (SC-GHG, or social cost of carbon), a tool to monetize the climate-related benefits of federal regulations and programs that would reduce GHG emissions. The withdrawn SC-GHG could also have been used to estimate the climate-related costs of revising or rescinding regulations that would increase GHG emissions. For additional discussion, see CRS In Focus IF10625, *Social Costs of Carbon/Greenhouse Gases: Issues for Congress*, by (name redacted).

Table 3. RIAs for Economically Significant CAA Rules, 2001-2016

Year	# of Economically Significant CAA Regs	# for which Estimated Benefits Exceeded Costs	# for which Estimated Costs Exceeded Benefits	# with Substantial Overlap of Cost and Benefit Ranges	# for which Costs or Benefits Were Undetermined
2001	0				
2002	1	1			
2003	1	1			
2004	5	3			2
2005	3	2	1		
2006	3	3			
2007	2	2			
2008	5	4		1	
2009	2	1			1
2010	7	7			
2011	7	6	1		
2012	5	5			
2013	1	1			
2014	1	1			
2015	6	4			2
2016	6	5			1
Total	55	46	2	1	6

Source: CRS, based on OMB, Office of Information and Regulatory Affairs (OIRA), *Historical Reports*.

CRS looked at the RIAs prepared by EPA under the G. W. Bush and Obama Administrations. From 2001-2016, EPA completed RIAs for 55 CAA rules under the executive order. Information concerning these rules is provided in Table 3.

In general, the agency concluded that the benefits of these rules would exceed the costs: 46 of the 55 RIAs reached this conclusion. Two rules, one promulgated in 2005 and the other in 2011, projected costs greater than benefits. In the other seven cases, either ranges of cost and benefit had a substantial overlap, or the agency was unable to quantify or monetize the costs or benefits.

In addition to E.O. 12866, the Clean Air Act itself, in Section 317, requires the EPA Administrator to prepare an economic impact assessment for several types of air quality standards, including

- section 111 new stationary source performance standards and regulations for existing stationary sources,
- title I, part C, prevention of significant deterioration standards,
- section 202, mobile source standards,
- section 211(c) fuel and fuel additive standards, and
- section 231, aircraft standards.[19]

[19] Section 317 also lists "title I, part B, ozone and stratosphere protection standards," among the standards for which economic impact assessments must be prepared, but title I, part B was repealed by the Clean Air Act Amendments of 1990, P.L. 101-549.

The assessment is to contain an analysis of (1) the costs of compliance with any such standard or regulation; (2) the potential inflationary or recessionary effects of the standard or regulation; (3) the effects on competition of the standard or regulation with respect to small business; (4) the effects of the standard or regulation on consumer costs; and (5) the effects of the standard or regulation on energy use.

The standards listed in section 317 generally mention or imply cost as a consideration in the statute; thus, the economic impact assessment is to inform that consideration. However, section 317 also states that "nothing in this section shall be construed to provide that the [assessment] affects or alters the factors which the Administrator is required to consider in taking any action [when promulgating or revising the listed standards]."

The section is limited in two other respects. In subsection (d), it gives the Administrator discretion to limit the time and resources devoted to the required analyses:

> The assessment required under this section shall be as extensive as practicable, in the judgment of the Administrator taking into account the time and resources available to the Environmental Protection Agency and other duties and authorities which the Administrator is required to carry out under this Act.

And in subsection (e), it provides that

> Nothing in this section shall be construed ... to authorize or require any judicial review of any such standard or regulation, or any stay or injunction of the proposal, promulgation, or effectiveness of such standard or regulation on the basis of failure to comply with this section.

Issues Raised by EPA's CAA Cost-Benefit Analyses

A number of issues have been raised regarding EPA's cost-benefit analyses for Clean Air Act rules. Four issues are discussed below.

Cumulative Impacts of Clean Air Act Regulations

A frequent criticism of EPA's Clean Air Act regulations is that the agency underestimates the cost and other negative impacts of rules by considering them individually, and thus potentially ignoring cumulative impacts. Other critics assert that, by considering rules individually, EPA cost-benefit analyses may double count the benefits of simultaneous regulations. EPA's RIAs do focus on individual rules, because both the statute—in the many places that it requires consideration of cost or economic factors—and

E.O. 12866 require the agency to weigh costs and economic factors and consider options for individual rules.

The agency starts RIAs with a baseline of state and federal regulatory requirements already promulgated. The RIA then estimates the additional costs and benefits of the proposed or final rule under consideration. In both proposed rule and final rule RIAs, the agency generally considers more stringent and less stringent options in order to provide analysis of the costs and benefits of each.

In some cases, there may be more than one rule addressing pollution from a specific industry under development simultaneously. This happens when the agency is implementing congressional directives found in different sections of the act. For example, in the past five years, fossil-fueled power plants have been the subject of rules addressing interstate transport of sulfur dioxide and nitrogen oxides (under Section 110(a)), emissions of hazardous air pollutants (under Section 112(d)), and emissions of the greenhouse gas carbon dioxide (under Sections 111(b) and (d)). As individual rules are promulgated and implemented, their requirements are added to the baseline, but when several rules are *proposed* simultaneously, each rule's impact must be examined in isolation in order to comply with statutory and executive order directives. Addressing the rules in isolation means that the cost-benefit analysis may ignore the cumulative economic impact of new regulations. It can also mean that adding the costs and benefits of simultaneous proposals might lead to double counting and over-estimating both costs and benefits.

Although developing required rules simultaneously may pose analytical issues, it has practical advantages for the regulated entities. It can allow a regulated facility to choose pollution control approaches that address several problems at once, e.g., installing a scrubber that will both reduce sulfur dioxide emissions and address hazardous air pollutants, or switching to cleaner fuels that eliminate or reduce the pollution problems addressed by all of the proposed regulations. It may also save on compliance costs by permitting the affected entities to address multiple regulations during a single outage.

EPA Studies of Cumulative Costs and Benefits

Although most cost-benefit analyses have focused on individual rules, EPA has conducted three analyses of the cumulative impact of Clean Air Act regulations, as required by Section 812 of the Clean Air Act Amendments of 1990.[20] Each of the three analyses found that the benefits of Clean Air Act regulations far exceed the cost.

---

[20] Section 812 of the 1990 amendments amended Section 312 of the underlying statute. Although the requirement is codified with Section 312 (at 42 U.S.C. 7612), the studies are generally referred to as the Section 812 studies. Section 812 required the Administrator, in consultation with the Secretary of Commerce, the Secretary of Labor, and the Council on Clean Air Compliance Analysis, to submit a report to Congress within 12 months of enactment summarizing all costs incurred previous to the enactment of the Clean Air Act Amendments of 1990 in the effort to comply with CAA standards, and all benefits that have accrued to the United States as a result of such costs. The section required updates to the report every 24 months thereafter; the updates were to make projections into the future regarding expected costs, benefits, and other effects of compliance with CAA standards, as well as updating the first report.

- The first of the studies, a retrospective study entitled, *The Benefits and Costs of the Clean Air Act, 1970 to 1990*, was completed in 1997. It estimated that the cumulative cost of Clean Air Act regulations between 1970 and 1990 was $523 billion (in 1990 dollars). The benefits of those regulations outweighed the costs by more than an order of magnitude, according to the agency. The estimated economic value of benefits ranged from $5.6 to $49.4 trillion over the 20-year period, depending upon the assumptions employed, with a mean value of $22.2 trillion. Human health effects accounted for the vast majority of this economic value: the agency concluded that the regulations reduced premature mortality by 205,000 persons annually.[21]

  The agency noted a number of limitations and uncertainties in the data. On the cost side, the agency noted that the estimate "... does not include several potentially important indirect costs which could not be readily quantified, such as the possible adverse effects of Clean Air Act implementation on capital formation and technological innovation."[22] On the benefit side, the agency noted, "... it is important to recognize the substantial controversies and uncertainties which pervade attempts to characterize adverse human health and ecological effects of pollution in dollar terms."[23] In addition, the estimates "
  ... do not include a number of other potentially important benefits which could not be readily quantified, such as ecosystem changes and air toxics-related human health effects."[24] Nevertheless, the agency concluded, "Given the magnitude of difference between the estimated benefits and costs, ... it is extremely unlikely that eliminating these uncertainties would invalidate the fundamental conclusion that the Clean Air Act's benefits to society have greatly exceeded its costs."[25]
- The second study, a prospective study entitled *The Benefits and Costs of the Clean Air Act, 1990 to 2010*, was released in November 1999. The study estimated the cost of compliance for regulations under the 1990 amendments to Titles I through V of the Clean Air Act at $19 billion annually in the year 2000 (in 1990$), rising to $27 billion annually in 2010. The estimated economic value of benefits ranged from $16 billion to $160 billion annually in 2000, and $26 billion to $270 billion in 2010. Although costs slightly exceeded benefits at the low end of the benefit estimate, EPA concluded that benefits exceeded cost by more than 4 to 1 for the

[21] See U.S. EPA, *The Benefits and Costs of the Clean Air Act, 1970 to 1990*, Report to Congress, October 1997, Table ES-1 and, more generally, pp. ES-2 to ES-9, at https://www.epa.gov/sites/production/files/2015-06/documents/contsetc.pdf.

[22] Ibid., Abstract, p. 2.

[23] Ibid., p. ES-7.

[24] Ibid., Abstract, p. 2.

[25] Ibid.

central estimate.[26] The study estimated costs and benefits separately for Title VI, which deals with protection of the stratospheric ozone layer. The benefits and costs for this title were estimated for a 175-year period, reflecting the slow nature of repairing the ozone layer. The agency estimated benefits of $530 billion over that time, with costs of $27 billion.[27]

- The third study, another prospective study, is entitled *The Benefits and Costs of the Clean Air Act, 1990 to 2020*. This study was released in March 2011. The study estimated the annual cost of compliance for regulations under the 1990 amendments to the Clean Air Act at approximately $65 billion in 2020, with a central estimate of benefits of $2 trillion. Using the central estimates, benefits exceed costs by 31 to 1. As with the earlier studies, "Most of these benefits (about 85%) are attributable to reductions in premature mortality associated with reductions in ambient particulate matter.... " The agency estimated that "cleaner air will ... prevent 230,000 cases of premature mortality" in 2020, at a cost of $280,000 per premature mortality avoided.[28]

OIRA Reports on the Cost and Benefit of Regulations

The Office of Information and Regulatory Affairs (OIRA) in the President's Office of Management and Budget (OMB) is the office that conducts interagency reviews of proposed and final regulations under E.O. 12866. In addition, OIRA prepares annual reports to Congress on the cost and benefit of regulations.[29]

The latest such final report, for 2015, includes estimates of the aggregated annual benefits and costs of regulations reviewed by OMB over the last 10 years.[30] In a section on "EPA Air Rules," the report states: "Across the Federal government, the rules with the highest estimated benefits as well as the highest estimated costs, by far, come from the Environmental Protection Agency and in particular its Office of Air and Radiation. Specifically, EPA rules account for 61 to 80 percent of the monetized benefits and 44 to 55 percent of the monetized costs."[31] The OMB report stated that EPA Office of Air rules in the 10-year period had benefits that were 4 to 21 times as great as their cost.[32]

---

[26] U.S. EPA, *The Benefits and Costs of the Clean Air Act, 1990 to 2010*, Report to Congress, November 1999, pp. ii-iv, at https://www.epa.gov/sites/production/files/2015-07/documents/fullrept.pdf. The report contains a discussion of the limitations and uncertainties of the data similar to that in the 1997 report.

[27] Ibid., p. v.

[28] U.S. EPA, *The Benefits and Costs of the Clean Air Act from 1990 to 2020*, Final Report, March 2011, Abstract.

[29] These reports are required under the Regulatory Right-to-Know Act (31 U.S.C. § 1105), which was enacted as part of the Treasury and General Government Appropriations Act for FY2001 (P.L. 106-554).

[30] Office of Management and Budget, *2015 Report to Congress on the Benefits and Costs of Federal Regulations and Agency Compliance with the Unfunded Mandates Reform Act*, at https://obamawhitehouse.archives.gov/sites/default/files/omb/inforeg/2015_cb/2015-cost_benefit-report.pdf.

[31] Ibid., p. 12.

[32] Ibid., Table 1-2.

The Role of Particulate Matter in EPA Regulatory Impact Analyses

A second criticism of EPA cost-benefit analyses is that the estimated benefits often rely on the effects of reducing a single category of pollutants, particulate matter (PM). Research has tied PM to tens of thousands of premature deaths, and EPA often finds that reductions in PM emissions justify regulation, even where the target of the regulations is a different pollutant.

In many of these cases, the RIAs do not monetize the benefits of controlling the emissions that were the primary target of the regulation. For example, an RIA that accompanied the 2004 National Emission Standards for Hazardous Air Pollutants from Industrial, Commercial, and Institutional Boilers and Process Heaters (the "2004 Boiler MACT") estimated that there would be $16 billion of annual benefits due to reductions in sulfur dioxide and particulate matter emissions. But it also stated

> This analysis does not quantify the benefits associated with reductions in hazardous air pollutants (HAP). The magnitude of the unquantified benefits associated with omitted categories and pollutants, such as avoided cancer cases, damage to ecosystems, or materials damage to industrial equipment and national monuments, is not known.[33]

Of the 22 EPA air rules considered in the 2015 OIRA report cited above, the highest estimated benefits were for three rules promulgated in 2005, 2007, and 2012.[34] For these rules, and others promulgated under the Clean Air Act, OIRA notes

> the large estimated benefits of EPA rules issued pursuant to the Clean Air Act are mostly attributable to the reduction in public exposure to fine particulate matter (referred to in many contexts as PM). While some of these rules monetize the estimated benefits of emissions controls designed specifically to limit particulate matter or its precursors, other rules monetize the benefits associated with ancillary reductions in particulate matter that come from reducing emission of hazardous air pollutants which are difficulty (sic) to quantify and monetize because of data limitations. For example, in the case of the Utility MACT, particulate matter "co-benefits," make up the majority of the monetized benefits, even though the regulation is designed to limit emissions of mercury and other hazardous air pollutants. The consideration of co-benefits, including the co-benefits associated with reduction of particulate matter, is consistent with standard accounting practices and has long been required under OMB Circular A-4.[35]

---

[33] U.S. EPA, *Regulatory Impact Analysis for the Industrial Boilers and Process Heaters NESHAP*, Final Report, February 2004, p. 10-1.

[34] The three rules are the Clean Air Fine Particle Implementation Rule issued in 2007, with benefits estimates ranging from $19 billion to $167 billion per year; the Clean Air Interstate Rule issued in 2005, with benefits estimates ranging from $12 billion to $152 billion; and the National Emission Standards for Hazardous Air Pollutants From Coal- and Oil-Fired Electric Utility Steam Generating Units ("Utility MACT" or "MATS" rule) issued in 2012, with benefits estimates ranging from $28 billion to $77 billion (all figures in 2001$).

[35] OMB, *2015 Report to Congress*, p. 13 [note omitted].

There are hundreds of air pollutants that Congress required or authorized EPA to regulate under the Clean Air Act. Congress directed EPA to set emission standards for sources of 187 hazardous air pollutants that are listed in the statute. Many of these are categories of pollutants (e.g., arsenic compounds, fine mineral fibers, polycyclic organic matter) rather than individual substances, so there are more than 187 pollutants to consider. Although there is research indicating that these pollutants are carcinogenic, mutagenic, teratogenic, neurotoxic, cause reproductive dysfunction, or are otherwise acutely or chronically toxic, in most cases there are not data regarding the concentrations to which populations are exposed, or epidemiological data regarding illness or mortality associated with exposure to the individual pollutant. The agency proceeds with regulation because it was directed by the statute to do so, but it may not be able to quantify or monetize the benefits of regulating emissions of a specific substance.

Defining "Particulates"

Particulate matter (also known as particle pollution, particulates, or PM) is a category of pollutants rather than a specific chemical. EPA identifies PM as "a complex mixture of extremely small particles and liquid droplets. Particle pollution is made up of a number of components, including acids (such as nitrates and sulfates), organic chemicals, metals, and soil or dust particles."[36] Hazardous air pollutants, if not particles themselves, often adhere to particles in the emissions. Because PM includes so many different pollutants, many of the regulations targeting hazardous air pollutants rely on technologies that capture PM. Given the broad nature of particulate emissions, most of the available pollution control technologies (scrubbers, fabric filters, electrostatic precipitators, carbon or other sorbent injection, use of catalysts, etc.) capture particulate emissions or PM precursors.[37]

The agency does, however, have an established, peer-reviewed methodology for estimating the benefits of reductions in emissions of particulate matter,[38] which have been linked to increased mortality in numerous scientific studies. Most air pollutants are particulates, and most EPA air quality regulations reduce particulate emissions, either as the targeted pollutant, or as a co-benefit of reducing emissions of some other pollutant.

---

[36] U.S. EPA, Office of Air and Radiation, "Particulate Matter," at http://www.epa.gov/pm/.

[37] The term "precursor" refers to a pollutant that reacts with other substances in the atmosphere to form another air pollutant. Sulfur dioxide (SO_2), for example, is a precursor of sulfate particles and sulfuric acid, both of which are considered particulates.

[38] For a discussion of this methodology, see, for example, U.S. EPA, *Regulatory Impact Analysis (RIA) for Residential Wood Heaters NSPS Revision*, Final Report, February 2015, pp. 7-1 to 7-9, at http://www2.epa.gov/sites/production/files/2015-02/documents/20150204-residen tial-wood-heaters-ria.pdf.

The Value of a Statistical Life[39]

Another reason that particulates play such an important role in RIAs is that they are linked to premature mortality. When premature mortality is avoided, the monetization of that benefit, using what is called "the value of a statistical life," generally is greater than the value of all other benefits combined.[40] This raises another issue: the role played by the methodology used to value lives saved.

The value of statistical lives saved (VSL) has played an important role in RIAs for many years. EPA adopted guidelines under President Reagan that, in updated form, have guided its VSL analyses since 1983. The guidelines were most recently updated in 2010.[41]

In general, the VSL is estimated by using one of two methodologies: "willingness to pay" (stated preference), or "willingness to accept" (revealed preference). The first of these methods uses surveys in which respondents are asked how much they would be willing to pay to avoid particular risks or outcomes. For example, if 100,000 people are each willing to pay an average of $50 to reduce a 1 in 100,000 risk of dying from exposure to a particular risk, then the value of a statistical life for the population relative to that risk is $5 million ($50 times 100,000). Revealed preference studies, on the other hand, use data from market transactions or observed behavior to estimate the value of certain risks. One example is wage-risk studies, in which researchers compare workers' earnings in occupations with varying levels of on-the-job risks.[42]

The Office of Management and Budget's Circular A-4, which more fully delineates the regulatory analysis requirements in Executive Order 12866, was "designed to assist analysts in the regulatory agencies by defining good regulatory analysis ... and standardizing the way benefits and costs of Federal regulatory actions are measured and reported."[43] It states that economists tend to view willingness-to-pay as "the most appropriate measure of opportunity costs,"[44] and that the willingness-to-pay approach is

---

[39] This section of this report, discussing the value of a statistical life, is adapted from archived CRS Report R41140, *How Agencies Monetize "Statistical Lives" Expected to Be Saved By Regulations*, by (name redacted). Curtis Copeland is no longer at CRS; questions about that report can be directed to (name redacted), Specialist in Government Organization and Management.

[40] Other benefits considered in Regulatory Impact Analyses include health benefits, such as the avoidance of nonfatal heart attacks, hospital and emergency room visits, cases of respiratory symptoms, cases of aggravated asthma, cases of chronic bronchitis, number of days when people miss work, and the number of days when people must restrict their activities. Environmental effects, including improvements in visibility in national parks, reductions in damage to ecosystems and building materials, and improvements in fishing, agricultural yields, and forest productivity, are also frequently identified as benefits of a rule in RIAs.

[41] U.S. Environmental Protection Agency, *Guidelines for Preparing Economic Analyses*, December 17, 2010, at https://yosemite.epa.gov/ee/epa/eerm.nsf/vwAN/EE-0568-51.pdf/$file/EE-0568-51.pdf.

[42] For additional information, see archived CRS Report R41140, *How Agencies Monetize "Statistical Lives" Expected to Be Saved By Regulations*, by (name redacted), or U.S. Office of Management and Budget, 2014 Report to Congress on the Benefits and Costs of Federal Regulations and Unfunded Mandates on State, Local, and Tribal Entities, p. 15, at https://obamawhitehouse.archives.gov/sites/default/files/omb/inforeg/2014_cb/2014-cost-benefitreport.pdf.

[43] OMB Circular A-4, "Regulatory Analysis," September 17, 2003, p. 1. The circular is available at https://obamawhitehouse.archives.gov/sites/default/files/omb/assets/regulatory_matters_pdf/a-4.pdf.

[44] Ibid., p. 20.

"the best methodology to use if reductions in fatality risks are monetized."[45] In monetizing health benefits, the circular states that a willingness-to-pay measure is "the conceptually appropriate measure as compared to other alternatives (e.g., cost of illness or lifetime earnings), in part because it attempts to capture pain and suffering and other quality-oflife effects," and also because it "allows you to directly compare your results to the other benefits and costs in your analysis."[46]

Released in 2003, the circular noted that academic studies had identified VSLs from $1 million to $10 million, but it did not recommend that agencies use any particular VSL.[47] A 2010 EPA guidance reported academic estimates of VSL ranging from $0.85 million to $19.8 million in 2006 dollars.[48] The guidance states that EPA uses a 1997 estimate of VSL, updated to current dollars by applying the Gross Domestic Product price deflator. In recent RIAs, the agency has used an estimate of $9.9 million in 2015 dollars.

The Difficulty of Quantifying Costs or Benefits

A fourth issue, recognized by both proponents and opponents of cost-benefit analysis, is the difficulty of quantifying both costs and benefits. Cost-benefit analysis is an imperfect tool that may fail to provide accurate projections in both cases. At least three factors contribute to or help explain the difficulty: a) the key role played by assumptions in making projections; b) the paucity of retrospective studies that might provide better methods or data; and c) particularly for benefits, the inability to quantify or monetize effects in light of existing information.

Assumptions

On the cost side, assumptions need to be made regarding the control technology or production methods that will be used to achieve compliance, and the costs of various inputs, such as energy, the price of which may be subject to substantial volatility. New technologies may encounter unforeseen implementation difficulties that result in cost overruns. On the other hand, control options have often benefitted from technology improvements or economies of scale that result in lower costs than predicted. Varying assumptions can lead to large differences between EPA's cost-benefit estimates and those of affected stakeholders.[49]

---

[45] Ibid., p. 29.

[46] Ibid., p. 28.

[47] Ibid., p. 30.

[48] U.S. Environmental Protection Agency, National Center for Environmental Economics, *Guidelines for Preparing Economic Analyses*, September 2000, available at http://yosemite.epa.gov/ee/epa/eed.nsf/pages/Guidelines.html. The discussion regarding valuation of mortality and morbidity risks is on pp. 87-98.

[49] See, for example, the discussion of industry and EPA estimates of the costs and benefits of the 2015 Ozone NAAQS in CRS Report R43092, *EPA's 2015 Ozone Air Quality Standards*, by (name redacted) and (name redacted).

Often the assumptions, whether made by stakeholders or by EPA, fail to foresee broad economic factors that end up determining how industry will comply with the standard being promulgated. For example, when EPA promulgated the Mercury and Air Toxics Standards (MATS) for electric power plants in 2012, the RIA concluded that coal-fired power would increase its share of total electric power, rising 14% from its 2009 level to 1,982 billion kilowatt-hours in 2015, while natural gas-fired power would decrease 16% to 710 billion kilowatt-hours. Given this reliance on coal-fired plants, the power sector would need to make an enormous investment in pollution control equipment to clean up their emissions.

A major factor in the RIA's analysis was the price of natural gas: the analysis concluded that natural gas would cost $5.32 per million Btu in 2015 after the rule took effect. (The price was expressed in 2007 dollars, which would be $6.07 in 2015 prices.) With gas at this price—and projected to increase further[50]—it would make sense to invest in keeping coal-fired power plants running. By 2015, however, the cost of natural gas had fallen to $2.63, and most analysts concluded that prices would remain low.[51] As a result, rather than spend the money to control emissions from coal-fired power plants (which the RIA estimated at $9.6 billion per year), many utilities found it easier and cheaper to retire coal-fired plants and increase the use of natural-gas-fired plants; in many cases, these plants were available and underutilized. The result was that coal-fired power declined to 1,352 billion kilowatt-hours (32% below the RIA projection) in 2015, and natural gas-fired power increased to 1,333 billion kilowatt-hours (88% above the RIA projection).

Retrospective Studies

The MATS rule RIA illustrates another issue related to the cost of regulations. While EPA and other agencies churn out dozens of RIAs annually as they develop regulations, there are few studies of the actual cost of rules once they've been implemented (what economists refer to as "ex post" costs, as opposed to the "ex ante" costs estimated in RIAs). EPA recognizes this issue. A 2014 agency study states:

> In 2010, then Deputy Administrator Bob Perciasepe inquired about research on retrospective cost analysis, particularly of past EPA regulations. An investigation of the literature revealed that the collection of retrospective analyses of EPA regulations is thin and no generalized conclusions could be drawn. Bob Perciasepe asked the National Center for Environmental Economics (NCEE) to design and launch a retrospective cost analysis with the goal of improving EPA's cost assessments.[52]

---

[50] In April 2012, the U.S. Energy Information Administration's *Annual Energy Outlook* (AEO 2012) projected reference case (Henry Hub) natural gas prices at $8.95 per MMBtu in 2030.

[51] In AEO 2015, the price was projected to remain below $6.00 through 2030. AEO 2016 projected prices below $5.00 as late as 2040.

[52] U.S. Environmental Protection Agency, National Center for Environmental Economics, *Retrospective Study of the Costs of EPA Regulations: A Report of Four Case Studies,* August 2014, Acknowledgements, at https://yosemite.epa.gov/ee/epa/eerm.nsf/vwan/ee-0575.pdf/$file/ee-0575.pdf.

The result was *Retrospective Study of the Costs of EPA Regulations: A Report of Four Case Studies*.[53] The report states: "The literature posits a number of hypotheses for why one might expect ex ante and ex post cost estimates to differ, yet ex post cost case studies are too few in number and narrow in scope to lend strong support for one hypothesis over another."[54] The report produced four case studies:

> For each case study, we assessed whether it would be possible to collect sufficient ex post compliance cost information using only publicly-accessible data sources. In general, we found that while data for some necessary components are readily available, the cost information is generally lacking. ... While several of the case studies are suggestive of overestimation of costs ex ante, we do not consider the current evidence to be conclusive.[55]

Quantification and Monetization

Most rules also have benefits that cannot be quantified or monetized in light of existing information. It is common for EPA to list in its RIAs benefits that it believes will result from a rule, but that it was unable to quantify or monetize. In the RIA for the Cross-State Air Pollution Update Rule, promulgated in October 2016, for example, EPA stated: "Data, time, and resource limitations prevented the EPA from quantifying the impacts to, or monetizing the co-benefits from several important benefit categories.... "[56] The agency listed seven categories of health benefits and 25 categories of welfare benefits that it did not quantify.

OMB's Circular A-4 recognizes these difficulties. It states:

> It will not always be possible to express in monetary units all of the important benefits and costs. When it is not, the most efficient alternative will not necessarily be the one with the largest quantified and monetized net-benefit estimate. In such cases, you should exercise professional judgment in determining how important the non-quantified benefits or costs may be in the context of the overall analysis.[57]

CONCLUSION

Although many parts of the Clean Air Act require the EPA Administrator to promulgate regulations without mentioning consideration of cost, EPA is bound by the

---

[53] Ibid.

[54] Ibid., p. vii. On page 4, the report summarizes five earlier studies by a variety of authors, who found that ex ante costs were often overestimated.

[55] Ibid., pp. vii-viii.

[56] U.S. Environmental Protection Agency, *Regulatory Impact Analysis of the Cross-State Air Pollution Rule (CSAPR) Update for the 2008 National Ambient Air Quality Standards for Ground-Level Ozone*, p. ES-16, at https://www3.epa.gov/ttn/ecas/docs/ria/transport_ria_final-csapr-update_2016-09.pdf.

[57] OMB Circular A-4, p. 2.

statute in some cases and by executive orders in the case of each economically significant rule to provide estimates of the costs and benefits during the rulemaking process.[58] The agency has indicated that benefits exceed costs, usually by a wide margin, for the vast majority of its CAA rules: as noted earlier, according to EPA, the estimated cumulative benefits of CAA regulations during the period 1990 2020 will exceed the estimated costs by more than 30 to 1.

Projecting that benefits will exceed costs may not be sufficient under the Trump Administration, however. The President has spoken repeatedly of the need to reduce the cost of regulation, which he believes has restrained the growth of the economy and "killed" jobs. On January 30, 2017, he signed Executive Order 13771, "Reducing Regulation and Controlling Regulatory Costs." A second Executive Order, E.O. 13783, "Promoting Energy Independence and Economic Growth," signed March 28, 2017, addressed specific Clean Air Act regulations.

The first of these two executive orders addressed regulations promulgated by all federal agencies. Press coverage focused on its requirement that "... whenever an executive department or agency ... publicly proposes for notice and comment or otherwise promulgates a new regulation, it shall identify at least two existing regulations to be repealed."[59] Other elements of the order are also worth noting:

- First, the executive order does not mention the benefits of regulation. It focuses exclusively on costs.
- Second, it establishes a process under which agencies shall be given an annual regulatory budget, with the Director of OMB identifying "a total amount of incremental costs that will be allowed for each agency in issuing new regulations and repealing regulations for the next fiscal year." For FY2017, the E.O. directs that "the total incremental cost of all new regulations, including repealed regulations, to be finalized this year shall be no greater than zero, unless otherwise required by law or consistent with advice provided in writing by the Director of the Office of Management and Budget.... "
- Third, it requires the Director of OMB to "provide the heads of agencies with guidance on the implementation of this section. Such guidance shall address, among other things, processes for standardizing the measurement and estimation of regulatory costs; standards for determining what qualifies as new and offsetting regulations; standards for determining the costs of existing regulations that are considered for elimination; processes for accounting for costs in different fiscal

[58] In addition, in 2015, the Supreme Court held that EPA is required to consider costs when determining if it is "appropriate and necessary" to regulate power plants under Section 112. *Michigan v. EPA*, 135 S. Ct. 2699 (2015). For additional information, see CRS Report R43699, *Key Historical Court Decisions Shaping EPA's Program Under the Clean Air Act*, by (name redacted) and (name redacted).

[59] "Reducing Regulation and Controlling Regulatory Costs," Executive Order 13771, January 30, 2017, Section 2.

years; methods to oversee the issuance of rules with costs offset by savings at different times or different agencies; and emergencies and other circumstances that might justify individual waivers of the requirements of this section." Guidance was issued on April 5, 2017.[60]

- Fourth, it states that, "Nothing in this order shall be construed to impair or otherwise affect ... the authority granted by law to an executive department or agency, or the head thereof.... "

How this order will affect CAA rules remains to be seen. It gives the OMB Director new authority in tasking him with the provision of guidance, the identification of regulatory budgets, and the discretion to grant waivers from the order's requirements. The April 5, 2017, guidance reinforces this, stating in numerous places that OMB (in the form of the Director of its Office of Information and Regulatory Affairs (OIRA)) will address issues "on a case-by-case basis."[61]

Some rules under the Clean Air Act might be exempt from the executive order's requirements. Whether this is the case will depend on what interpretation is given to the order's language exempting from the regulatory budgets regulations that are "otherwise required by law." The authority granted by law to the EPA Administrator in at least 25 sections or subsections of the CAA (identified in Table 2) directs the Administrator to set or review standards without subjecting that authority to cost considerations. In many other cases—whether or not the CAA allows consideration of costs—binding deadlines for EPA rulemaking have been established by the courts. The OMB guidance describes such cases as "judicially required rulemaking," and includes in this category rules for which deadlines have been established by a settlement agreement or consent decree.[62] This might cover a substantial number of CAA rules.[63]

The second of the two orders, E.O. 13783, requires reviews of all agency actions "that potentially burden the development of domestically produced energy resources, with

[60] Executive Office of the President, Office of Management and Budget, "Guidance Implementing Executive Order 13771, Titled 'Reducing Regulation and Controlling Regulatory Costs,'" Memorandum from Dominic J. Mancini, Acting Administrator, Office of Information and Regulatory Affairs, April 5, 2017, 17 p., at https://www.whitehouse.gov/sites/whitehouse.gov/files/omb/memoranda/2017/M-17-21.pdf.

[61] Ibid. See the responses to questions 16, 17, 18, 19, 22, 27, 28, and 37.

[62] Ibid., Q6, p. 5.

[63] It is not entirely clear what OIRA will do in such cases: the guidance states that statutorily or judicially required actions "may qualify for a full or partial exemption from EO 13771's requirements." (Answer to Q33) Later in the same Answer, it states that "EO 13771 does not prevent agencies from issuing regulatory actions in order to comply with an imminent statutory or judicial deadline, even if they are not able to satisfy EO 13771's requirements by the time of issuance. However, agencies will be required to offset any such EO 13771 regulatory actions as soon as practicable thereafter." Elsewhere, in discussing regulatory actions in which the law prohibits the consideration of costs in determining a statutorily required standard, the guidance states, "Because EO 13771 applies only to the extent permitted by law, agencies are still required to comply with their statutory obligations. Accordingly, if a statute prohibits consideration of cost in taking a particular regulatory action, EO 13771 does not change the agency's obligations under that statute. However, agencies will generally be required to offset the costs of such regulatory actions through other deregulatory actions taken pursuant to statutes that do not prohibit consideration of costs. Because each agency's obligations will differ depending on the particular statutory language at issue, these issues must be addressed on a case-by-case basis." (Answer to Q18).

particular attention to oil, natural gas, coal, and nuclear energy resources."[64] The order addresses specific CAA regulations, including the Clean Power Plan for existing fossil-fueled electric generating units (EGUs) and two proposed rules related to it, the New Source Performance Standards for new and modified EGUs, and the New Source Performance Standards for the Oil and Natural Gas Sector. Each of these rules would control emissions of greenhouse gases from an energy-producing sector. The E.O. directs EPA to review these rules "for consistency with the policy set forth in section 1 of this order," and, if appropriate, to "suspend, revise, or rescind" them.

Section 1 lists many goals, including to

- "promote clean and safe development of our nation's vast energy resources,"
- "ensure that the Nation's electricity is affordable, reliable, safe, secure, and clean,"
- "take appropriate actions to promote clean air and clean water," and
- ensure that "necessary and appropriate environmental regulations comply with the law, are of greater benefit than cost, when permissible, achieve environmental improvements for the American people, and ... employ the best-available peer-reviewed science and economics."

Most of the commentary on this executive order has presumed that the purpose of the reviews will be to rescind the rules in question; but the rules were originally justified by EPA as measures that would achieve most of Section 1's listed goals. Rescinding the rules would likely require a new justification that can withstand judicial scrutiny. Thus, "review" of the rules does not automatically equate with "rescind."[65]

Whatever becomes of these specific rules, EPA's administration of the Clean Air Act under the Trump Administration is likely to function differently than it did in the Obama Administration. The CAA gives the EPA Administrator broad discretion in deciding whether regulations are necessary and how stringent they should be: many sections of the act, even some that don't allow consideration of cost, include phrases such as "in the judgment of the Administrator" or "as determined by the Administrator." Such language would seem to allow the Administrator a measure of discretion that will continue to shape Clean Air Act and other EPA regulations—both in deciding on the stringency of new regulations and in deciding whether new regulations are warranted. Unless this language is modified, it will also continue to provide fertile ground for legal arguments regarding the power that Congress delegated to EPA when it fashioned the various authorities that the act provides.

---

[64] Executive Order 13783, "Promoting Energy Independence and Economic Growth," March 28, 2017, Section 2.

[65] For a discussion of the required procedures for amending or rescinding rules, see CRS Report R41546, *A Brief Overview of Rulemaking and Judicial Review*, by (name redacted).

In: The Clean Air Act and The Clean Power Plan
Editor: Claire Ferguson

ISBN: 978-1-53614-825-1

Chapter 3

CLEAN AIR PERMITTING: IMPLEMENTATION AND ISSUES*

Claudia Copeland

SUMMARY

The 1990 Clean Air Act (CAA) amendments required major industrial sources of air pollutants to obtain operating permits. These permits, authorized in Title V of the act, are intended to enhance environmental compliance by detailing for each covered facility all of the emission control requirements to which it is subject. Title V also was intended to generate permit fees that would be used by state and local permitting authorities for administering the program. Implementation of these requirements affects more than 15,000 industrial sources of air emissions, as well as state and local air pollution control agencies. Adding these provisions to the act was controversial, and implementation, too, has generated controversies.

The Environmental Protection Agency (EPA) issued regulations to implement Title V in 1992. Aspects of those rules (particularly concerning procedures to modify permits) have been contentious since then. EPA has considered a number of regulatory revisions but has not finalized any modifications. However, EPA has issued white papers and a number of formal and informal guidance documents that, together with the 1992 rules, comprise the agency's current interpretation of statutory and regulatory requirements.

Because of regulatory and program approval delays, state and local agencies were slow to begin issuing Title V permits, falling far short of statutory deadlines and EPA's goals. According to an EPA Inspector General report, key factors that delayed issuance of permits included insufficient state resources, complex EPA rules and limited guidance, and conflicting state priorities. Now, however, most initial permits have been issued, and

* This is an edited, reformatted and augmented version of a Congressional Research Service report, Publication No. RL33632, dated September 1, 2016.

permit reissuance (required after five years) and modification have replaced issuance of initial permits as the major ongoing task of permitting agencies.

Attention to the Title V program increased in 2010 when EPA initiated several controversial regulatory actions to regulate emissions of greenhouse gases (GHGs) under existing CAA authority with implications for Title V permits. For Title V, these actions mean including GHG control requirements in Title V permits issued for non-GHG. To minimize the costs and administrative burden of its GHG regulations, EPA issued a "Tailoring Rule" to impose requirements only on the largest sources of GHG emissions. In June 2014, the Supreme Court found that EPA exceeded its statutory authority in issuing the Tailoring Rule but upheld the agency's authority to require sources that already need permits for conventional pollutant emissions to comply with CAA requirements for GHGs.

Most stakeholders agree that at least some of the benefits of Title V identified by Congress in the 1990 CAA amendments have been achieved, such as incorporation of applicable air pollution control requirements in a single document that is accessible to regulators, the public, and industrial sources. At the same time, there also has been widespread dissatisfaction with the program's complexity, costs, and confusing requirements. Many believe that a lack of EPA guidance and oversight has contributed greatly to implementation problems. Congressional examination of Title V has been limited to a few oversight hearings, but none recently. Clean Air Act issues have been of considerable interest during the 114th Congress, especially scrutiny of EPA's regulation of greenhouse gas emissions. Although EPA's actions concerning GHGs involve multiple provisions of the act, this congressional attention has not included Title V.

INTRODUCTION

A key element of the 1990 Clean Air Act (CAA) amendments (P.L. 101-549; 42 U.S.C. §§7661- 7661f) was the comprehensive permit program established in Title V of the act. It was added to the CAA to enhance compliance by detailing for each covered facility all the emission control requirements to which the facility is subject. Previously, an industrial source's pollution control obligations—ranging from emission controls and monitoring to record-keeping and reporting requirements—were scattered throughout numerous, often hard-to-find provisions of state plans or various federal regulations. While approximately 35 states or localities had operating permit programs before 1990, they varied considerably, and few were as comprehensive as the new Title V program. Adding these provisions to the CAA was controversial, and implementation, too, has generated controversies.

Congress's intent in creating the operating permits program was to "(1) better enforce the requirements of the law by applying them more clearly to individual sources and allowing better tracking of compliance, and (2) provide an expedited process for implementing new control requirements." Benefits of the air permit program were expected to include clarification of pollution control requirements, simplification of procedures for modifying a source's control obligations, augmenting state resources through permit fees, and enhancing states' ability to administer other significant new CAA responsibilities, such as the air toxics and acid deposition programs.[1]

This chapter describes the statutory background of the Title V program and the status of implementation, in terms of federal approval of state and local permitting authorities' programs and permit issuance. Regulatory actions by the Environmental Protection Agency (EPA) to limit emissions of greenhouse gases that will have implications for Title V permits are reviewed. It also discusses broad policy issues identified by various stakeholders, including program complexity, costs and permit fees, and inconsistencies due to a lack of sufficient federal guidance.

MAJOR FEATURES OF THE OPERATING PERMITS PROGRAM

Prior to the 1990 amendments, the CAA required individual permits only for construction of new or modified industrial sources of air pollutants. Existing sources did not have to obtain permits unless they were subsequently modified and increased their air emissions. Title V, which was modeled after a similar program in the Clean Water Act, expanded the number of sources requiring federal permits by stipulating that all major pollution sources and other designated sources must obtain operating permits, which permitting agencies will use to ensure compliance with the CAA.

Operating permits are now an important tool in the overall compliance and enforcement aspects of the CAA, because permits contain the plans and schedules for sources to reach attainment with provisions of the act, plus emission limitations and monitoring requirements. Further, they are enforceable, meaning that EPA, states, or citizens may take action to require a source to achieve compliance with the terms of its permit.

EPA estimates that nearly 15,500 industrial sources are subject to Title V. These specifically include major sources, defined in the CAA as stationary facilities that emit or have the potential to emit 100 tons or more per year of any regulated pollutant or combination of pollutants, and sources subject to the act's acid rain provisions. Title V also covers:

- sources in nonattainment areas[2] that emit as little as 10 tons per year of volatile organic compounds (VOCs), depending on the region's nonattainment status;

---

[1] U.S. Congress, Senate, Committee on Environment and Public Works, "Clean Air Act Amendments of 1989, report to accompany S. 1630," S.Rept. 101-228, 101st Congress, 1st session, pp. 346-348.

[2] Nonattainment areas are regions that have failed to attain and maintain one or more National Ambient Air Quality Standard (NAAQS). EPA has promulgated NAAQS for six air pollutants: sulfur dioxide, particulate matter, nitrogen dioxide, carbon monoxide, ozone, and lead. For background information, see CRS Report RL30853, *Clean Air Act: A Summary of the Act and Its Major Requirements*, by James E. McCarthy and Claudia Copeland.

- sources subject to New Source Performance Standards (NSPS);[3]
- regulated sources of air toxics emissions (any source that emits more than 10 tons per year of an individual hazardous pollutant or more than 25 tons per year of any combination); and
- sources required to have new source or modification permits under Title I of the act.

Permits compile in a single document all of the enforceable emission limitations and standards, plus inspection, monitoring, compliance certification, and reporting requirements for the source, but they are not intended to change or alter the existing, underlying requirements or add any substantive requirements. Permits generally contain these elements: emissions limitations and standards to assure compliance with all applicable requirements; monitoring, recordkeeping, and reporting; fee payments; and an annual certification by a responsible official of the source. They are issued for five-year periods and must be renewed thereafter.

Title V is intended to be primarily a state-run program. It provides two incentives in this regard: (1) permitting agencies are able to use permit fees collected from sources to run their permit programs and (2) EPA will implement a federal operating program if a state fails to do so. In many cases, states delegate to local programs the responsibility for implementing the operating permits programs. For example, 34 local authorities implement the permit program in California, rather than a single state agency. In other locations (for example, Tennessee, Arizona, Washington, and Idaho), local authorities implement the program in parts of a state, and the state regulatory agency has responsibility elsewhere.

Permitting agencies are required to collect permit fees sufficient to cover the cost of the permit program, and the fees may be used only for administering the program. The act requires that the fee schedule be set so as to collect from all sources, in the aggregate, not less than $25 per ton of each regulated pollutant to cover all reasonable (direct and indirect) costs of administering the program.[4] Regulated pollutants include VOCs, National Ambient Air Quality Standard pollutants except carbon monoxide, and pollutants regulated under the hazardous air pollutant and NSPS provisions of the act.

---

[3] New Source Performance Standards establish nationally uniform, technology-based maximum emission levels for categories of new major stationary sources, such as power plants, steel mills, etc. The goal is to require the installation of new pollution control technology and thus prevent new pollution problems.

[4] The fee amount considered sufficient to fund all permit program costs (sometimes called the "presumptive minimum") has been adjusted for inflation by EPA annually from the original level of $25 per ton of regulated pollutant. The current inflation-adjusted presumptive minimum amount is $48.49 (Sept. 1, 2015-Aug. 31, 2016). States are not required to adopt the presumptive minimum fee rate; they may charge higher or lower fees, so long as they collect sufficient fees to cover their administrative expenses.

EPA Regulations

EPA issued regulations to implement the permit program requirements in July 1992 (40 C.F.R. Part 70), seven months after the statutory deadline. The Part 70 rules cover the minimum elements of state permit programs. Rules promulgated in July 1996 and amended in 1999 (40 C.F.R. Part 71) cover federal permitting on Indian reservations and in any state that fails to adopt or implement an approvable Title V program.

While state and local agencies primarily implement the Title V program, the CAA provides for EPA oversight of permitting programs and gives EPA the right to review permit applications and object to proposed Title V permits. Defining the parameters of this oversight has been a source of tension among EPA, states, and industry. One overarching issue has been the question of developing permits with sufficient flexibility to allow for the fact that industrial sources often change their operations (thus affecting air emissions) in response to marketplace signals, and they want to be able to do so quickly without needing a revised permit for every such change. The public also has an important role in the permitting process, because citizens may submit comments and request a hearing on draft permits. Further, under CAA Section 505(b), any person may petition EPA to object to issuance of a permit, and Section 502(b)(6) requires that there be an opportunity for judicial review in state court of final permit actions. Judicial review may be sought by the applicant, persons who participated in the public comment process, and any other person who could obtain judicial review of such actions under state laws.

The Part 70 regulations were challenged by environmentalists, states, and industry, mostly over details concerning permit revisions. In response, in 1994, EPA proposed modifications intended to address some of the biggest issues. Those proposed changes were not widely accepted by states and industry, however, and EPA took additional steps to streamline the permit programs. Since 1994, EPA has worked with states, industry, and environmentalists to reach consensus on key issues and has issued and re-drafted several regulatory proposals but has not finalized major modifications to Part 70.

The focus of efforts to revise the rules has been on how to streamline aspects of the regulations concerning permit modifications and revisions. According to EPA officials, the 1992 regulations give states ample flexibility to write permits that allow for minor operational changes by industry and to process permit modifications. Still, efforts to revise those rules have been driven by concerns of many permitting authorities and industrial sources that the current rules do not allow sufficient flexibility. One issue that has been particularly contentious concerns the procedures and amount of public review required for relatively minor modifications at emissions sources. At issue have been the definition and criteria for facility changes that are neither so minor that little or no review will be required, nor clearly so environmentally significant that full EPA review and public comment will be appropriate. EPA continued internal and external discussions on revising the Part 70 rules for some time, but by early 2005, agency officials apparently had ceased these activities.

However, in 2009, the Obama Administration finalized a rule announced as one of the Bush Administration's final regulatory actions. Called the Flexible Air Permit rule, it clarifies existing regulations concerning how industry can make changes to facilities without having to obtain a new permit. It is intended to reaffirm opportunities for accessing operational flexibility under current regulations, while ensuring existing levels of environmental protection.[5]

During the lengthy consideration of possible Part 70 rule changes, EPA issued guidance in the form of white papers to address some implementation issues that had been raised, such as permit revisions, EPA's role in reviewing permits, administrative complexity, etc. The first was issued in 1995 (concerning streamlining to reduce costs and paperwork), and the second in 1996 (concerning overlapping federal and state requirements). A third white paper, concerning options for operational flexibility, was drafted in mid-2000, but was criticized by environmentalists and some state officials and was not issued (instead, EPA issued the flexible air permit rule in 2009, as discussed above). In addition, EPA has issued some formal guidance on specific implementation issues.[6]

STATUS OF THE PERMIT PROGRAM AND PERMIT ISSUANCE

Under the CAA, all states and territories were to submit operating permit programs by November 15, 1993. Once a state or locality's program was approved by EPA (within one year), major industrial sources had one year to submit permit applications to the state or local authority, and permitting agencies had three years to issue permits.

As of June 1997, EPA had approved permit programs for all 114 submissions by states, local agencies, and territories. In part because of regulatory and program approval delays, state and local agencies were slow to begin issuing Title V permits. In early 1998, EPA surveyed states and EPA regions and found that only about 2,100 permits had been issued, compared with an estimated 11,000 that should have been issued. In March 1999, EPA proposed an ambitious goal of resolving the permit backlog by the end of 2000. Permit issuance increased, but it continued to fall far short of the statutory deadlines and EPA's goals. As of March 2001, 57% had been issued. While there were potential consequences for permitting authorities that did not get program approval by EPA (i.e., the possibility of federal takeover, which EPA initiated in a few cases but did not impose), there were no real consequences or penalties for industrial sources that did not receive Title V permits

[5] U.S. Environmental Protection Agency, "Operating Permits Program; Flexible Air Permit Rule," 74 *Federal Register* 51418-51440, October 6, 2009.

[6] EPA maintains a website that contains a database of over 230 Title V guidance and policy documents that help to interpret the Part 70 and Part 71 rules. See https://www.epa.gov/title-v-operating-permits/title-v-operating-permitpolicy-and-guidance-document-index.

due to state and local agencies' delays. The principal consequence was a delay in attaining the policy objectives intended by Congress when it established the Title V program.

A 2002 EPA Inspector General (IG) report criticized EPA and state and local agencies over the program's continuing problems. The IG noted that, as of December 31, 2001, 30% of required permits had not been issued, and only 4 state and 17 local agencies had issued all of their Title V permits. Key factors delaying the issuance of permits included insufficient state resources, complex EPA regulations and limited guidance, and conflicting state priorities, the IG reported.[7] A consequence of those delays was that the benefits that Congress intended the permit program to achieve have not been realized, according to the IG's analysis. According to EPA, as of January 2008—12 years after the first program approvals—99% of all original permits required for Title V sources had been issued. After that initial process, the ongoing tasks of permit reissuance (required after five years) and modification have become the focus of most permitting authorities' attention. According to EPA, nationally, every year about 100 new sources are required to obtain initial permits, and about 3,000 sources are required to obtain renewal permits.

Implications of Regulating Greenhouse Gas Emissions for Title V Permitting

In 2010 EPA initiated several CAA regulatory actions to limit emissions of greenhouse gases (GHGs) having implications for Title V permits and permitting. The actions followed from the agency's issuance in December 2009 of an "endangerment finding" under Section 202 of the act, which permits (in fact, requires) EPA for the first time to regulate pollutants for their effects as greenhouse gases. Relying on this finding, EPA finalized GHG emission standards for new cars and light trucks on April 1, 2010. The implementation of these standards, in turn, triggered CAA permitting requirements and the imposition of technology-based control requirements for new and modified major stationary sources of GHGs (e.g., power plants and industrial facilities) beginning in January 2011.[8] Affected facilities are subject to the permitting requirements of Title V and the Prevention of Significant Deterioration (PSD) provisions of the law. For PSD, this includes state determinations of what constitutes Best Available Control Technology (BACT) that affected facilities will be required to install to limit GHG emissions. For Title V, this means including GHG control requirements in Title V permits.

As noted above, under Title V, major stationary sources are defined as those that emit or have the potential to emit more than 100 tons or more per year of any air pollutant subject to EPA regulation. For greenhouse gases, this is a relatively low threshold: EPA initially

---

[7] U.S. Environmental Protection Agency, Office of the Inspector General, "Evaluation Report: EPA and State Progress in Issuing Title V Permits," Report No. 2002-P-00008, March 29, 2002, pp. 8-14.

[8] For information, see CRS Report R40506, *Cars, Trucks, and Climate: EPA Regulation of Greenhouse Gases from Mobile Sources*, by James E. McCarthy and Brent D. Yacobucci.

general public. While many believe that the permit program has clarified requirements and fostered consistency and fairness in regulatory treatment of sources, others argue otherwise, saying that permits suffer from excessive length and increased complexity. Benefits are more likely to be observed in states and localities that had no operating permit program prior to Title V and are likely to be questioned more vigorously where such a program previously did exist. Likewise, views are mixed on whether the Title V program has resulted in air quality and health benefits. Arguably, this was not an objective, because Title V is an administrative program and was not intended to have a direct impact on emissions. Permitting agencies, however, say that many major sources have voluntarily restricted their operating conditions or installed pollution controls in order to reduce emissions below the Title V regulatory thresholds (thus becoming what is often referred to as "synthetic minors"; see discussion on page 10), which is a plus for the environment.

Critiques of the Title V program are reflected in a 2005 report of the EPA Inspector General[16] and in a 2006 report by a Task Force on Title V Implementation Experience that was convened by EPA's Clean Air Act Advisory Committee. The Task Force, consisting of state and local permitting agency, industry, and environmental advocacy group representatives, developed an extensive list of recommendations for program improvements, which it believed could be implemented under current legislative and regulatory authority.[17]

Beginning in 2003, EPA regional offices began conducting periodic audits of state, local, and territorial permitting programs to identify good practices, areas needing improvement, and ways that EPA can improve its oversight. Many but not all of these evaluations are available on EPA websites.[18] The evaluations identify a range of issues, such as timely permit issuance a nd backlogs, public participation, permit quality and documentation, staffing and training needs, adequacy of permit fee revenues, and communication with EPA and the public.

Complexity

Prior to establishment of the Title V program, major sources of air pollution were not required to have federally enforceable operating permits. Regulatory requirements for these

---

[16] U.S. Environmental Protection Agency, Office of the Inspector General, "Substantial Changes Needed in Implementation and Oversight of Title V Permits If Program Goals Are to Be Fully Realized," Report No. 2005-P00010. Hereinafter, 2005 EPA IG Report.

[17] The Title V Task Force, "Final Report to the Clean Air Act Advisory Committee on the Title V Implementation Experience," April 2006, available at https://www.epa.gov/caaac/final-report-clean-air-act-advisory-committee-title-vimplementation-experience. Hereinafter, Title V Task Force Final Report.

[18] See, for example, evaluations for Region 2 (https://www.epa.gov/caa-permitting/title-v-program-evaluations-region2), Region 4 (https://www.epa.gov/caa-permitting/caa-permitting-epas-southeastern-region#Program Evaluations), Region 5 (https://www.epa. gov/caa-permitting/caa-permitting-epas-great-lakes-region), and Region 10 (https:// www. epa.gov/caa-permitting/permit-program-reviews-epa-region-10).

sources typically derive from multiple provisions of the CAA and were often scattered among multiple documents, thus complicating efforts to determine compliance and to provide effective enforcement of the law. Consolidating existing requirements (some dating from the 1970s) into a single, comprehensive document, while undoubtedly beneficial, also has resulted in permits that are lengthy and detailed, making it difficult to read them without precise knowledge of the individual source's operations, according to some stakeholders.

Many issues have arisen, including what level of detail must be incorporated in a Title V permit (versus streamlining or simplifying the permit), how small and insignificant emissions sources should be treated, and how newly issued CAA requirements such as hazardous air pollutant standards will be incorporated. Another issue is the extent to which permitting agencies can adapt or update a source's existing requirements (particularly those that were established many years earlier) so as to reflect its current operating conditions, without resulting in changes that exceed the scope of Title V, since the permit program was intended to consolidate but not substantively alter regulatory requirements. Title V was not intended to affect the stringency of requirements incorporated into a permit, but stakeholders disagree on whether actions such as the addition of new monitoring or new compliance methods affect stringency or are sometimes tantamount to creating new substantive requirements.

A related issue concerns whether a Title V permit should include a compliance schedule for facilities that have received notices of violation regarding applicable requirements identified in the permit. EPA regulations provide that if a source is not in compliance with a particular requirement, the source must provide in its permit application a description of how it will come into compliance and a schedule that includes a compliance plan. Environmentalists and EPA have disagreed over whether the permit can include compliance schedules over alleged—rather than settled—violations. Activists argue that permits must include compliance schedules to address allegations of permit violations (essentially making the permit an active mechanism to enforce permitted limits), while EPA, states, and industry contend that that goes beyond the scope of the permit. Federal courts have split on the issue. The Obama Administration reportedly supports inclusion of compliance schedules to correct alleged violations only when challengers can unambiguously demonstrate that the violations are occurring at the facility.[19]

From the beginning of the program, a key concern for EPA, states, regulated industries, and environmental advocacy groups has been the issue of operational flexibility in operating permits, the concept of allowing for operational change at a facility, while assuring compliance with applicable requirements and ensuring environmental protection. EPA addressed the issue in the 1992 Part 70 regulations and subsequent guidance (especially a 2000 draft white paper). Also, beginning in 1993, EPA sponsored a flexible permitting pilot

[19] "DOJ Explains Rationale for High Bar on Title V Compliance Schedules," *InsideEPA.com*, December 11, 2009.

program with industrial facilities in 13 states in an effort to evaluate opportunities to design air permits to accommodate operational flexibility. EPA evaluated the economic and administrative benefits of flexible permits and documented several environmental performance benefits (including emissions reductions). Based on its assessment of the pilots, in 2009, EPA finalized a Flexible Air Permit rule that clarified opportunities under existing regulations to allow industry to be market-responsive while ensuring equal or greater environmental protection than that achieved by conventional permits.[20]

Title V Permit Review Requests and Petitions

As described previously, the CAA requires permitting authorities to submit a proposed Title V permit to EPA for review, and the Administrator has 45 days to object to its issuance. If the Administrator does not object during the 45-day period review period, any person may petition the Administrator within 60 days after expiration of the 45-day period with specific objections. Rules to implement these aspects of the statute were promulgated in the 1992 Part 70 regulations. Since then, however, EPA has concluded that the permit review and petition process is not functioning well because it lacks transparency and consistency, and as a result, in August 2016, the agency proposed rule changes that are intended to streamline the process.[21] In the proposal, EPA noted that "over time, petitions have raised increasingly more complex policy, legal, and technical matters," and that the process has "evolved into a resource-intensive effort" by EPA.

Evidence of problems with the process includes the number of petitions that are backlogged and awaiting a decision by the agency. EPA maintains an online database of such petitions,[22] and as of August 2016, the database identified 149 pending petitions. Significantly, despite the CAA requirement that the Administrator is to grant or deny a petition within 60 days after it is filed, 143 of these petitions (96%) have been pending for more than 60 days, and 43 (30% of those pending more than 60 days) have been pending for more than a decade. Concerns about the permit review and petition process were raised as issues in the 2006 Title V Task Force report.

Among other things, the August 2016 proposal would spell out mandatory content and format for petitions, would require state and other permitting authorities to respond in writing to major concerns raised during the public comment period on draft Title V permits, and would encourage challengers to file their petitions electronically. Some elements of the proposal have previously been detailed in responses to individual petitions—such as agency interpretation of the statute— and the proposed new regulatory language also would codify those prior administrative orders. By establishing clearer procedural parameters,

---

[20] U.S. Environmental Protection Agency, "Operating Permits Program; Flexible Air Permit Rule," 74 *Federal Register* 51418-51440, October 6, 2009.

[21] U.S. Environmental Protection Agency, "Revisions to the Petition Provisions of the Title V Permitting Program," 81 *Federal Register* 57822-57846, August 24, 2016.

[22] See https://www.epa.gov/title-v-operating-permits/title-v-petition-database.

EPA intends that the changes will benefit permitting authorities, permitted sources, and potential petitioners. With the proposed changes, EPA expects that its ability to meet its statutory obligations to review proposed permits, respond to Title V petitions, and provide more transparency to the petition process will all be improved.

Costs and Permit Fees

Industry groups generally believe that the regulatory burden and costs of the Title V program outweigh the benefits and far exceed EPA's estimate at the time that the Part 70 rules were adopted of $15,000 average cost per facility annually. In addition to permit applications, facilities incur recurring costs related to staffing, permit changes/corrections, report preparation, legal reviews, and management reviews of compliance. Some stakeholders (environmental advocates, for example) argue that program cost is not a significant issue when viewed in the context of cost as a percentage of a company's operating cost, and that companies often benefit from the additional information gained through the program.

Costs are a continuing concern for state and local regulatory authorities with regard to increasing paperwork requirements and the sufficiency of current permit fees. Having adequate resources to administer federal environmental programs is always an issue for states and localities, especially in light of increasing program demands. State environmental agencies' revenue sources vary, but they generally depend on a combination of EPA and state grant funding—both of which have been in decline—Title V permit fees, and other fees, in a few cases. The Title V permit fee requirement was intended to ensure that sufficient resources would be provided for necessary permitting activities. EPA has not conducted a comprehensive audit of permitting agency resources, but available information suggests that a number of states (perhaps many) are not collecting sufficient fees to cover their costs, which contributes to problems in hiring staff and processing permits.

In 2007, an environmental advocacy group, the Environmental Integrity Project, released a report concluding that more than half of the states have fee structures that do not meet federal minimum standards. The report focused on 18 states in which fees fell below the federal presumptive minimum, either because the states set lower emission fees or set a ceiling on the amount that could be collected from each polluter that was lower than the federal presumptive minimum on at least the first 4,000 tons of emissions of each pollutant covered by a Title V permit. The report found that, if states were to raise fees to at least the minimum federal amount, they could provide a significant source of funding to support their air quality management programs, but that additional funding (for example, from EPA

CAA grants) also is needed. It recommended that EPA undertake a comprehensive evaluation to ensure that low emission fees are not weakening the CAA permit program or its enforcement.[23]

A 2011 survey of state air quality officials found that dwindling budgets are pushing some states to consider returning their delegated CAA permitting programs to EPA, unless they can raise permit fees charged to industry or identify other sources of revenue. States have authority to determine the level of fees that they want to set—EPA's "presumptive minimum" fee level is not mandatory—and the state survey found that they vary widely.[24]

One factor affecting fee revenues is the large number of sources that have lawfully opted out of the Title V program. EPA originally estimated that about 37,000 facilities nationwide would be subject to Title V. Today, the universe of Title V sources is nearly 15,500, and the others—often termed "synthetic minors"—have installed pollution control equipment or taken other steps to ensure that their emissions are below the Title V threshold of what is a major source. For permitting agencies, the resource issue is twofold. First, most agencies initially calculated fee amounts and revenues based on a larger number of sources, but actual revenues have been less, since more than one-half of sources have lawfully avoided Title V coverage. Second, these sources still represent a regulatory workload, in terms of non-Title V permitting (separate state requirements[25]) and tracking, which is unlikely to be covered fully by other fee revenues or available funding. Getting state legislatures to approve fee increases to address these concerns is politically difficult, in most cases.

As sources install pollution control equipment that reduces emissions, another issue arises. The majority of permitting authorities have fee structures that are based on emissions, such as a per-ton fee. Over time, revenues decrease because sources implement tighter emission controls and/or sources close. With decreasing emissions and reduced permit fee revenues, permitting authorities' ability to cover program costs and carry out required program activities is strained. The adequacy of permit fee revenues has been highlighted as a concern in a number of EPA's evaluations of state, local, and territorial permit programs, described above.

The EPA's Inspector General 2002 report identified insufficient permit fee resources as a key factor that caused delays in issuing Title V permits. A 2014 EPA IG report focused specifically on revenues to operate the program.[26] The IG found that states' permitting

[23] Environmental Integrity Project, "Shortchanging the Clean Air Act: An Analysis of State Revenues Lost Due to Low Emission Fees," March 2007.

[24] National Association of Clean Air Agencies, "NACAA Title V Fees Survey—State by State Results," September 2011.

[25] A source that voluntarily reduces its emissions or potential to emit below major source permitting thresholds, in order to opt out of Title V or other CAA permitting requirements, may seek a state operating permit from its permitting authority. State operating permits are issued at the discretion of the permitting authority. Because such a permit contains restrictions that underlie why it is avoiding Title V, it is referred to as a federally enforceable state operating permit (FESOP), and EPA may take enforcement action for violations.

[26] U.S. EPA Office of Inspector General, "Enhanced EPA Oversight Needed to Address Risks from Declining Clean Air Act Title V Revenues," Report No. 15-P-0006, October 20, 2014.

revenues have been declining in recent years and that state operating expenses often exceeded Title V revenues. Among nine of the largest state and local permitting authorities that oversee 45% of the nation's active Title V permits, there was a $69 million shortfall out of $672 million in expenses incurred between 2008 and 2012. Further, the IG criticized EPA for insufficient oversight of state permit fees and attributed the problem to several factors: a lack of a national oversight strategy, outdated (1993) guidance on fee collection, a lack of accounting expertise among EPA staff, and an unwillingness by some regions to pursue formal corrective actions against states. The report found, "The agency's weaknesses in identifying and obtaining corrective actions for Title V revenue sufficiency and accounting practices, coupled with declining resources for some permitting authorities, jeopardizes state and local Title V program implementation." The IG recommended a series of steps to improve the program, all of which EPA accepted, but EPA said that it will not complete most actions until the end of FY2017.

Confusing Requirements and Limited EPA Guidance

Critiques of the Title V program, including the EPA Inspector General's reports and the Title V Task Force report, identify insufficient EPA guidance as a major implementation issue. The IG's 2005 report, based on a review of permits issued by several states, found that permit clarity varies widely from state to state. Permit requirements are often vaguely stated or identified by reference to other documents, without narrative description or precise citation. EPA regulations require that the permitting authority prepare a statement for each draft permit to set forth the legal and factual basis of the permit, but the IG found that such statements often are inadequate or totally missing.

From its review, the IG concluded that many of these problems stem from a lack of EPA guidance on key issues, which results in permitting inconsistencies and contributes to permitting delays that adversely affect sources, agencies, and the public. The IG strongly recommended that EPA issue nationwide guidance on a number of topics (including requirements for the statement of basis in permits, requirements for content of annual compliance certifications by corporate officials, and applicability of sanctions for unresolved program deficiencies) in order to achieve more national cohesion and consistency in Title V permits.

EPA has issued very limited formal guidance and rules on Title V in the past several years. In lieu of formal guidance, EPA has relied on responses to citizen petitions and letters to regions and permitting authorities to convey its position on key Title V issues, arguing that this strategy allows for flexibility that reflects differences in source complexity.[27] Some stakeholders are concerned that, rather than resolving program issues through rule-making or nationally applicable interpretive guidance, EPA is using the petition process and other mechanisms not just to apply law to facts, but to make law in the first instance. The change away from issuing national rules or guidance occurred following a federal court decision which vacated a 1998 EPA guidance document on periodic monitoring. In that case, the court ruled that the agency had exceeded its authority by utilizing nonbinding guidance rather than national rule-making to interpret regulatory requirements.[28] The IG noted that permitting authorities may be unwilling to follow or may be unaware of guidance that isn't national in scope.

One example of lack of EPA guidance relates to incorporation of monitoring requirements in permits, one of the most contentious issues in Title V implementation. The CAA mandates that permits include monitoring and reporting requirements to assure compliance with permit terms and conditions. In the 1992 Part 70 rules, EPA required that permits include all monitoring and test methods detailed in the applicable underlying requirements (e.g., in NSPS or hazardous air pollutant standards incorporated in the Title V permit). In addition, where an underlying applicable requirement does not require "periodic monitoring," Part 70 requires that periodic monitoring be specified in the Title V permit.

The 2005 EPA IG report found that specification of monitoring requirements was one of the most significant areas of inconsistency in permits and that a number of stakeholders and EPA regional officials contend that more EPA guidance on these issues is needed. The Title V monitoring rules, and specifically the application of "periodic monitoring" requirements in permits, have been subject to several EPA interpretations and to litigation challenging those interpretations, in cases asking the federal court to determine the meaning of the periodic monitoring rule and related regulatory language and to determine the consistency of the rules (as interpreted by EPA) with the statutory requirements. In 2008, a federal appeals court vacated a 2006 rule as inconsistent with the Clean Air Act.[29] The 2006 rule had said that (contrary to prior policy) state and local authorities could not require supplemental monitoring in cases where existing monitoring is inadequate to ensure compliance with the Clean Air Act. According to the 2006 Title V Task Force report, stakeholders fundamentally disagree on the statutory and regulatory requirements and particularly on whether permitting authorities may or must (depending on one's perspective) specify new or revised emissions monitoring requirements in permits.[30] EPA officials said that a revised periodic monitoring rule would be developed, but none has been proposed.

[27] 2005 EPA IG Report, pp. 84-85.

[28] *Appalachian Power Co. v. EPA*, 208 F.3d 1015 (D.C. Circuit 2000).

[29] *Sierra Club v. EPA*, 536 F.3d 673 (DC Circuit 2008).

[30] Title V Task Force Final Report, pp. 47-63.

CONGRESSIONAL INTEREST

The Title V Task Force noted in 2006 that much about Title V remains unsettled and subject to debate in both the legal and policy arenas—an assessment that likely still applies—and it urged that steps be taken to "stem the tide of transaction costs and to bring an increased level of certainty and stability to implementation of this program."[31] Responding to the Task Force report in September 2006, EPA officials identified a number of priorities and next steps. Over the following one to two years, the agency said that it would work with states to identify best practices such as standards for granting a public hearing, providing written responses to public comments, and coordination of Title I and Title V process. Similarly, over the next one to three years, EPA would review existing guidance and issue new guidance as appropriate concerning clarification of the permit revision process and permit reopening, and clarification of certain statements required in draft permits. Finally, EPA would initiate rule-makings to address several concerns, such as clarifying when insignificant activities are exempt from permits, and allowing alternatives to newspapers for public notice requirements. Rulemakings could take three years or more, EPA said. Other priorities included improving online notice and online access to documents on the EPA website.[32] However, it appears that little follow-up activity has occurred.

In 2008, EPA officials reportedly told an industry group that the agency was working on a proposed rule and guidance document, in partial response to the Task Force report. The streamlining proposal and guidance were expected to address a number of issues—but only a small percentage of the Task Force's recommendations—including the types of emission units that must be included in a permit, the effectiveness of the public notice process, and the complexity and length of permits. The proposed rule was also expected to remove insignificant emission units from the Title V permit process, allow the use of online public notices instead of newspaper notices, and clarify the use of administrative amendments and minor permit revisions. A proposed rule was expected to be released in early 2009, but this has not occurred. In 2012, EPA officials indicated plans to make certain rule changes and issue guidance on how permit writers should include "statements of basis" in the enforceable terms of permits. The impetus for these plans, officials said, came from the 2005 EPA Inspector General report and President Obama's executive orders on reducing regulatory burdens.[33] No proposal has been issued.

Congressional oversight of the Title V program has been limited to hearings by a House Energy and Commerce subcommittee and a Senate Environment and Public Works

---

[31] Title V Task Force Final Report, pp. 8-9.

[32] "EPA's Response to the Title V Task Force Recommendations," presentation to the Clean Air Act Advisory Committee, September 14, 2006.

[33] Executive Order 13563, "Improving Regulation and Regulatory Review," 76 *Federal Register* 3821-3823, January 21, 2011, and Executive Order 13610, "Identifying and Reducing Regulatory Burdens," 77 *Federal Register* 28469- 28470, May 14, 2012.

consider technologies that will be fairly projected in the future so long as the technology is not speculative.

- *New source review* in areas not subject to NAAQSs, holding that EPA may override a state's determination of the "best available control technology" required for new stationary sources. EPA may require new source review for greenhouse gas emitters only if the new source will emit certain pollutants above threshold amounts.
- *The "routine maintenance" exemption* from NSPSs and new source review, created by EPA and accepted by the courts despite statutory silence. Courts hold that whether the exemption applies depends on the increase in a plant's expected life due to the project, and the project's cost, nature, and magnitude. Expansive interpretation of the exemption has been judicially rejected.
- *The "bubble concept,"* an EPA approach that looks at *net* changes in the emissions of a pollutant from a facility, holding that its permissibility depends on statutory context.
- *National standards for hazardous air pollutants*, holding that EPA may determine if a facility triggers the Act's "maximum achievable control technology" requirement for such pollutants by aggregating emission sources in a contiguous plant under common control, not just sources within the same source category. EPA is not limited in setting emission standards to hazardous air pollutants currently controlled with technology.
- *Greenhouse gas emissions*, holding that the CAA generally covers them, and that EPA has to exercise that authority based on policy concerns. See, however, "new source review" above.
- *Enforcement*, holding that the recipient of an administrative compliance order must be allowed to seek pre-enforcement review of the order in court.

GLOSSARY OF ACRONYMS

ACO	administrative compliance order
BACT	best available control technology
BSER	best system of emission reduction
CAA	Clean Air Act
CAIR	Clean Air Interstate Rule
CO_2	carbon dioxide
CPP	Clean Power Plan
EPA	Environmental Protection Agency
GHG	greenhouse gas
HAP	hazardous air pollutant
MACT	maximum achievable control technology
NAAQS	national ambient air quality standard
NESHAP	national emission standard for hazardous air pollutant
NO_x	nitrogen oxides
NEPA	National Environmental Policy Act
NSPS	new source performance standard
PSD	prevention of significant deterioration
RMRR	routine maintenance, repair, and replacement

SIP state implementation plan
SO_2 sulfur dioxide

Introduction

This chapter provides a selective overview of court decisions that have shaped the U.S. Environmental Protection Agency's (EPA) implementation of the Clean Air Act (CAA or Act).[2] The subjects these decisions cover are broad, because almost every major EPA rule has been challenged in court. Challenges to EPA actions are filed largely in the U.S. Court of Appeals for the D.C. Circuit (D.C. Circuit), as the CAA requires for challenges to "nationally applicable regulations."[3] Claims that EPA has failed to perform a nondiscretionary duty under the Act, such as meeting a statutory deadline, are challenged through "citizen suits" filed in federal district courts.[4] Over the 46 years of the modern CAA's existence, more than a dozen circuit and district court decisions on the Act have yielded Supreme Court opinions on appeal.

This chapter also highlights several court cases currently being litigated that address a number of unresolved questions related to EPA's CAA authority.

National Ambient Air Quality Standards

The central construct of the CAA is the *national ambient air quality standard* (NAAQS), a maximum concentration for an air pollutant that all areas of the country must meet. To establish NAAQSs, CAA Section 108 directs EPA to issue *air quality criteria*—documents assessing the scientific evidence on a pollutant's effects—for pollutants that "may reasonably be anticipated to endanger public health or welfare," and "the presence of which in ambient air results from numerous or diverse mobile or stationary sources ...."[5] Section 109 then requires EPA to set *primary NAAQSs* at a level requisite to protect public health, allowing an "adequate margin of safety," and *secondary NAAQSs* at a level requisite to protect the "public welfare" from any known or anticipated adverse effects.[6] Finally,

---

[2] 42 U.S.C. §§ 7401-7671q.

[3] CAA § 307(b); 42 U.S.C. § 7607(b). In contrast, petitions for review of EPA actions that are only "locally or regionally applicable" must be filed in the U.S. Court of Appeals for the "appropriate" circuit, though even these actions must be filed in the D.C. Circuit if "based on a determination of nationwide scope or effect." *Id.*

[4] CAA § 304(a); 42 U.S.C. § 7604(a).

[5] CAA § 108(a)(1); 42 U.S.C. § 7408(a)(1).

[6] CAA § 109(b); 42 U.S.C. § 7409(b). "Welfare" is a term of art, defined in the CAA to include at a minimum "effects on soils, water, crops, vegetation, manmade materials, animals, wildlife, weather, visibility, and climate, damage to and deterioration of property, and hazards to transportation, as well as effects on economic values and on personal comfort and well-being ...." CAA § 302(h); 42 U.S.C. § 7602(h).

Section 110 requires states to adopt *state implementation plans* to attain or maintain each NAAQS (see following section).[7]

In the 1970s, EPA established NAAQSs for six pollutants, commonly referred to as "criteria" pollutants.[8] The agency has not added any pollutants to the list since then, although it is required to review the existing standards at five-year intervals and promulgate revisions if appropriate.[9] These reviews have raised continuing issues among stakeholders, including states, industrial and other sources of pollution, and nongovernmental organizations (NGOs) in the health and environmental fields, particularly over the role of cost and feasibility in determining the level at which standards should be set. In directing EPA to set and revise the NAAQSs, Sections 108 and 109 make no reference to the cost or feasibility of attaining the standards—in contrast to other CAA sections where cost, feasibility, or the role of demonstrated technology are specifically required to be considered.

The silence of Sections 108 and 109 as to cost and feasibility led to a number of cases, all of which held that EPA must set NAAQSs without regard to their cost or technological feasibility. In *Lead Industries Ass'n v. EPA*, the D.C. Circuit addressed the primary and secondary NAAQSs for lead,[10] holding that requiring EPA to consider cost and feasibility in setting NAAQSs was "totally without merit."[11] Specifically rejected was the argument that the "adequate margin of safety" required by Section 109 to be factored into primary NAAQSs requires EPA to consider costs and feasibility. Persuasive to the court was that the CAA expressly provides for EPA to factor in costs and feasibility, and that even if the technology to achieve a NAAQS did not currently exist, the CAA was of a "technology forcing" character.[12]

In 2001, in *Whitman v. American Trucking Ass'ns*,[13] the Supreme Court unanimously affirmed *Lead Industries* finding it impermissible to consider costs in setting primary NAAQSs. The *Whitman* challengers had argued that the high costs of imposing a stringent primary NAAQS might injure health by closing down whole industries and impoverishing workers and consumers dependent on them. The *Whitman* challengers contended that EPA had to consider this effect on health, too. The Court disagreed, pointing out that CAA Section 110(f)(1) allowed EPA to waive compliance deadlines for stationary sources in certain circumstances, and that, as noted in *Lead Industries*, numerous CAA provisions allow EPA to take compliance costs into account.[14] That being so, the Court refused to

[7] CAA § 110(a)(1); 42 U.S.C. § 7410(1)(1).

[8] NAAQSs have been set for ozone, particulate matter, carbon monoxide, sulfur dioxide, nitrogen dioxide, and lead. 40 C.F.R. pt. 50.

[9] CAA § 108(d)(1); 42 U.S.C. § 7408(d)(1).

[10] 647 F.2d 1130 (D.C. Cir. 1980).

[11] *Id.* at 1148.

[12] *Id.* at 1149.

[13] 531 U.S. 457 (2001).

[14] For example, the Court noted CAA Section 202(a)(2), specifying that motor vehicle emission standards can take effect only "after such period as the [EPA] Administrator finds necessary to permit the development and application of the requisite technology, giving appropriate consideration to the cost of compliance ...." 42 U.S.C. § 7521(a)(2).

infer from the ambiguous language of Section 109 an authority for EPA to consider costs that elsewhere in the Act had been granted expressly.

State Implementation Plans

The CAA directs EPA to translate NAAQSs into emission limits for specific stationary sources. After EPA promulgated or revises NAAQS, CAA Section 110 requires each state to submit a *state implementation plan* (SIP) to EPA within three years.[15] SIPs specify what mix of federal, state, and local air pollution control measures the state will implement in order to reach or maintain the NAAQSs. To be approved by EPA, the SIP must satisfy a long list of requirements in Section 110(a)(2).[16] For example, the SIP must contain enforceable emission limitations, timetables for compliance, air quality monitoring, provisions addressing interstate pollution, and so on.[17]

Because Section 110(a)(2) does not require that SIPs be economically and technologically feasible, the Supreme Court concluded in *Union Electric Co. v. EPA* that EPA may not consider infeasibility when approving or disapproving a SIP.[18] Further, the Supreme Court held that states may submit SIPs more stringent than the CAA requires and EPA must approve them if they meet the Section 110(a)(2) factors.[19] In addition, as held in *Michigan v. EPA*, EPA may not condition SIP approval on a state's adoption of any specific control measure under the CAA.[20]

Finally, under *Indiana & Michigan Electric Co.*, EPA does not have to prepare an environmental impact statement under the National Environmental Policy Act (NEPA) when approving or disapproving a submitted SIP.[21] Courts have held EPA actions generally to be exempt from NEPA requirements, because the agency's mission of protecting the environment duplicates that of NEPA.[22]

Interstate Pollution

After EPA sets or revises a NAAQS, the governor of each state, using available monitoring data, must submit to EPA a list identifying each *air quality control region* in the state as either attainment, nonattainment, or unclassifiable for the pollutant in

[15] CAA § 110(a)(1); 42 U.S.C. § 7410(a)(1).
[16] 42 U.S.C. § 7410(a)(2).
[17] *Id.*
[18] 427 U.S. 246, 256 (1976). Several CAA provisions noted in the *Union Electric* decision have been amended since, but it has never been suggested that these amendments undermine the Court's essential holdings.
[19] *Id.* at 265.
[20] 213 F.3d 663, 687 (D.C. Cir. 2000).
[21] 509 F.3d at 842.
[22] *See, e.g.*, *Getty Oil Co. v. Ruckelshaus*, 467 F.2d 349, 359 (3d Cir. 1972).

question.[23] After reviewing these submissions, and often after negotiating over the boundaries of the listed areas, EPA formally promulgates a list of *nonattainment areas*.[24] Affected states are then required to submit SIPs or SIP revisions specifying what mix of federal, state, and local air pollution control measures will be implemented for each of the areas to reach or maintain the NAAQS.[25]

In many states, particularly in the Northeast, air quality is so affected by emissions from other states (referred to as "transported" air pollution) that it is difficult or impossible for the state to demonstrate how all areas in the state will reach attainment. To assist these downwind states, the CAA contains several provisions dealing with transported air pollution. The most important (and most frequently litigated) is Section 110(a)(2)(D),[26] the so-called *"good neighbor" provision.* Section 110(a)(2)(D)(i)(I) requires that a state prohibit stationary sources within the state from emitting air pollutants in amounts that will "contribute significantly" to NAAQS nonattainment, or "interfere with maintenance" of a NAAQS, in any other state. States must include such measures in the SIPs they submit to EPA.[27] When EPA determines that existing SIPs must be revised to satisfy the good neighbor provision (or other CAA requirements), it issues a "SIP call," sometimes to many states at once.[28]

Litigation over the good neighbor provision generally has involved emissions of nitrogen oxides (NO_x), particulates, and sulfur dioxide (SO_2), all of which can remain in the atmosphere and travel long distances from the point of emission. These substances are not only pollutants in their own right; they also contribute (through atmospheric reactions) to the formation of ozone, particulates, acid deposition, and other widespread regional pollution problems.[29] Litigation over the good neighbor provision has addressed how EPA defines when a state "contribute[s] significantly" to another state's pollution; what role cost-effectiveness can play in identifying control measures; whether regional cap-and-trade systems can substitute for emission control requirements imposed at specific sources; and the respective roles of EPA and the states in identifying required controls.[30]

---

[23] CAA § 107(d)(1)(A); 42 U.S.C. § 7407(d)(1)(A). For a list of air quality control regions, see 40 C.F.R. pt. 81.

[24] CAA § 107(d)(1)(B); 42 U.S.C. § 7407(d)(1)(B).

[25] CAA § 110(a)(1)-(2); 42 U.S.C. § 7410(a)(1)-(2).

[26] CAA § 7410(a)(2)(D).

[27] CAA § 110(a)(2); 42 U.S.C. § 7410(a)(2).

[28] CAA § 110(k)(5); 42 U.S.C. § 7410(k)(5). Besides Section 110(a)(2)(D), the CAA also addresses interstate pollution through Section 126, 42 U.S.C. § 7426. Section 126(b) allows a state or political subdivision to petition EPA to make a "finding that any major source or group of stationary sources emits or would emit any air pollutant in violation of the prohibition of section 110(a)(2)(D)(ii)." If EPA makes the finding, the source or sources in question cannot be constructed or can no longer operate, except under emission limitations and a compliance schedule prescribed by EPA.
Supplementing these provisions, CAA Section 115 creates a mechanism for dealing with emissions in the United States that might endanger health or welfare in a foreign country. 42 U.S.C. § 7415.

[29] *See* EPA, National Air Quality: Status and Trends of Key Air Pollutants, at https://www. epa.gov/air-trends.

[30] *See, e.g., North Carolina v. EPA* and *Michigan v. EPA* decisions discussed in immediately following paragraphs.

In 2000, the D.C. Circuit held in *Michigan v. EPA*[31] that nothing in Section 110(a)(2)(D) bars EPA from considering costs.[32] Consequently, EPA had acted lawfully when it determined that a state would no longer be contributing "significantly" to a downwind state's NAAQS nonattainment if the state reduced the relevant emissions by the amount that could be eliminated using "highly cost-effective controls" (those that eliminate a ton of the relevant pollutant for less than $2,000 per ton).[33] The court found that EPA may apply this standard uniformly to all the covered states, no matter the amount of each state's contribution.[34]

North Carolina v. EPA[35] involved a challenge to EPA's Clean Air Interstate Rule (CAIR), issued in 2005.[36] CAIR sought to reduce SO_2 and NO_x precursor emissions in 28 states in order to reduce nonattainment of the NAAQSs for fine particulate matter and ozone in downwind states. The D.C. Circuit, however, found CAIR to be flawed.[37] Most important to the court, CAIR's emissions trading program, though aimed at reducing emission-control costs as approved in *Michigan*, did not assure some "measurable" emission reduction in *each* upwind state.[38] Emissions reduction by the upwind states *collectively* was not enough to satisfy Section 110(a)(2)(D).[39] Second, the court said, EPA must give independent effect to the "interfere with maintenance" prohibition in Section 110(a)(2)(D)—not, as CAIR did, as a prohibition triggered only after the "contribute significantly" prohibition was triggered.[40]

EPA's effort to remedy the CAIR deficiencies identified in *North Carolina* led it to issue the Cross-State Air Pollution Rule, also known as the Transport Rule or CSAPR, in 2011.[41] This rule addresses the same pollutants and the same states as did CAIR. Under CSAPR, an upwind state is deemed to "contribute significantly" to downwind nonattainment to the extent its pollution produces more than 1% of the NAAQS concentration in at least one downwind state and could be eliminated cost-effectively. Through modelling, EPA set the total amount of pollution an upwind state could produce in a given year. This second EPA effort also was found by the D.C. Circuit to be inconsistent with Section 110(a)(2)(D),[42] but this time the decision was reversed by the Supreme Court. In *EPA v. EME Homer City Generation, L.P.*, the High Court upheld

---

[31] 213 F.3d 663 (D.C. Cir. 2000).
[32] *Id.* at 679.
[33] *Id.* at 680.
[34] *Id.* at 679-680.
[35] 531 F.3d 896 (D.C. Cir. 2008).
[36] 70 Fed. Reg. 25,162 (May 12, 2005).
[37] Though initially vacating the CAIR rule, the D.C. Circuit changed its mind some months later, leaving CAIR in effect "until it is replaced by a rule consistent with our opinion." *North Carolina v. EPA*, 550 F.3d 1176, 1178 (D.C. Cir. 2008).
[38] 531 F.3d at 908.
[39] *Id.*
[40] *Id.* at 910.
[41] 76 Fed. Reg. 48,208 (Aug. 8, 2011).
[42] 696 F.3d 7 (D.C. Cir. 2012).

CSAPR.[43] First, the Court found no fault with the fact that EPA, having found the relevant SIPs to be inadequate, had along with CSAPR promulgated federal implementation plans allocating each state's total allowed emission amount among sources of that pollutant within the state. The CAA, held the Court, does not require that states be given a second opportunity to file a SIP after EPA has quantified the state's emissions budget.[44] Second, the Court held that nothing in the Good Neighbor Provision compelled the "cost-blind" interpretation of the D.C. Circuit.[45] Rather, it concluded, EPA's allocation of emission reductions among upwind states based on cost-effectiveness is a permissible, workable, and equitable reading of the Provision.[46]

NEW SOURCE PERFORMANCE STANDARDS

CAA Section 111[47] directs EPA to develop federal "standards of performance" for new, modified, and reconstructed stationary sources of air pollution, called *New Source Performance Standards* (NSPSs). NSPSs are nationally uniform, lessening the incentive for companies to "shop" for locations with less stringent requirements. "Standards of performance" are defined as emission standards reflecting the amount of emission reduction "achievable" through the use of the "best system of emission reduction" (BSER) that is "adequately demonstrated," "taking into account" cost and nonair quality impacts.[48]

Most of the phrases just quoted have been litigated, beginning in the CAA's early years. Case law holds that "adequately demonstrated" does not necessarily imply that any existing source of the type proposed for an NSPS is able to meet the NSPS. Rather, *Portland Cement Ass'n v. Ruckelshaus* says that Section 111 "looks toward what may fairly be projected for the regulated future, rather than the state of the art at present ...."[49] Still, *Lignite Energy Council v. EPA* cautions that "EPA may not base its determination that a technology is adequately demonstrated or that a standard is achievable on mere speculation or conjecture ...."[50] The agency may compensate for the absence of emissions data in a new source category by, for example, "extrapolation of a technology's performance in other industries."[51] Where EPA *is* able to show that existing sources of the type proposed for an NSPS can meet the NSPS, *National Lime Ass'n v. EPA* instructs that those existing sources must be representative of the industry as a whole.[52] Finally, courts have noted that Section 111 is

---

43 134 S. Ct. 1584 (2014).
44 *Id.* at 1600-1602.
45 *Id.* at 1583.
46 *Id.* at 1603-1609.
47 42 U.S.C. § 7411.
48 CAA § 111(a)(1); 42 U.S.C. § 7411(a)(1).
49 486 F.2d 375, 391 (D.C. Cir. 1973).
50 198 F.3d 930, 934 (D.C. Cir. 1999).
51 *Id.*
52 627 F.2d 416, 432-433 (D.C. Cir. 1980).

silent as to the weight to be given the cost and nonair quality impacts of a control technology, the section saying only that they shall be "taken into account."[53] Given that silence, *Lignite Energy Council* and other D.C. Circuit decisions have granted EPA much discretion: "EPA's choice [of best adequately demonstrated technology] will be sustained unless the environmental or economic costs of using the technology are exorbitant."[54]

As to what standards are "achievable," *Essex Chemical Corp. v. Ruckelshaus* describes an achievable standard as "within the realm of the adequately demonstrated system's efficiency and which, while not at a level that is purely theoretical or experimental, need not necessarily be routinely achieved within the industry prior to its adoption."[55] The question on what standards are considered "adequately demonstrated" is central in the litigation over EPA's regulations limiting CO_2 emissions from new power plants.[56] Twenty-five states—led by North Dakota and West Virginia—have filed petitions in the D.C. Circuit challenging EPA's final NSPS for carbon dioxide (CO2) emissions from new, modified, and reconstructed fossil fuel-fired power plants under CAA Section 111(b), which it calls the "Carbon Pollution Standards."[57] The states have been joined by other petitioners including a labor union, a rural electric cooperative association, several other fossil-fuel-related companies and utilities, and several industry and trade groups.[58] One of the primary issues raised by the petitioners is whether the technologies on which EPA based the standards of performance, including carbon capture and sequestration/storage (CCS), have been "adequately demonstrated" or are the "best system" under Section 111(b).[59] A three-judge panel is to hear oral argument for this case on April 17, 2017.[60]

Emission Guidelines under CAA Section 111(d)

At the same time or after it issues Section 111(b) NSPSs, EPA must establish, under certain circumstances, emission guidelines for existing sources in that category pursuant to Section 111(d).[61] These guidelines establish binding requirements that states are required to address when they develop plans to regulate the existing sources in their jurisdictions.[62]

---

[53] *See, e.g., New York v. Reilly*, 969 F.2d 1147, 1150 (D.C. Cir. 1992).

[54] *Lignite Energy Council*, 198 F.3d at 933.

[55] 486 F.2d 427, 433-434 (D.C. Cir. 1973).

[56] *North Dakota, et al. v. EPA*, No. 15-1381 (D.C. Cir. filed Oct. 23, 2015); Standards of Performance for Greenhouse Gas Emissions From New, Modified, and Reconstructed Stationary Sources: Electric Utility Generating Units; Final Rule, 80 Fed. Reg. 64,510 (Oct. 23, 2015) (hereinafter "NSPS Final Rule").

[57] *See generally* docket for *North Dakota v. EPA*, No. 15-1381 (D.C. Cir. filed Oct. 23, 2015); Standards of Performance for Greenhouse Gas Emissions From New, Modified, and Reconstructed Stationary Sources: Electric Utility Generating Units; Final Rule, 80 Fed. Reg. 64,510 (Oct. 23, 2015) [hereinafter *NSPS Final Rule*].

[58] For further information on the legal challenge to the NSPS Final Rule, see CRS Report R44480, *Clean Power Plan: Legal Background and Pending Litigation in West Virginia v. EPA*, by Linda Tsang and Alexandra M. Wyatt.

[59] *See, e.g.*, State Pet'rs' Opening Br., *North Dakota v. EPA*, Nos. 15-1381 (D.C. Cir. filed Oct. 13, 2016).

[60] Order, *North Dakota v. EPA*, No. 15-1381 (D.C. Cir. Dec. 2, 2016).

[61] CAA Section 111; 42 U.S.C. § 7411(d)(1).

[62] *Id.*

Similar to Section 110 of the CAA, which requires states to develop and revise implementation plans to achieve EPA's NAAQS and subsequent changes to those standards, Section 111(d) directs EPA to establish state plan "procedures."[63]

On the same day that EPA issued the Carbon Pollution Standards for new, existing, and reconstructed power plants under CAA Section 111(b), the agency finalized emission guidelines to regulate CO_2 emissions from existing fossil-fuel power plants under Section 111(d). These guidelines, commonly referred to as the Clean Power Plan (CPP),[64] have been among the more controversial environmental regulations that EPA has ever promulgated, as reflected by the multiparty litigation over the CPP in the D.C. Circuit.[65] One of the key issues in the litigation is whether EPA has the authority under the CAA to go beyond the "fenceline" of an individual regulated source and consider shifting from fossil fuel to natural gas-fired power and renewable energy in the interconnected electric grid to establish the BSER.[66] Another issue in the CPP litigation relates to the interpretation of two differing amendments of Section 111(d)(1)(A) that were enacted into law, which sets forth exclusions to EPA's authority to issue Section 111(d) emission guideline rules.[67]

On September 27, 2016, the en banc (full court) D.C. Circuit heard oral argument in the CPP litigation.[68] Judicial decisions in this and the litigation over the Carbon Pollution Standards for new power plants will likely establish the boundaries of EPA's authority under CAA Section 111.

NEW SOURCE REVIEW

The CAA seeks not only attainment of NAAQSs in dirty-air areas; it also aims to limit air quality deterioration in areas that have met or exceeded the NAAQSs. To achieve the latter goal, the Act's *Prevention of Significant Deterioration* (PSD) program imposes requirements in addition to those in Section 111 for new and newly modified stationary

---

[63] *Id.*; CAA Section 110; 42 U.S.C. § 7410.

[64] Carbon Pollution Emission Guidelines for Existing Stationary Sources: Electric Utility Generating Units, Final Rule, 80 Fed. Reg. 64,661, 64,663 (October 23, 2015). For additional background, see CRS Report R44341, *EPA's Clean Power Plan for Existing Power Plants: Frequently Asked Questions*, by James E. McCarthy et al.

[65] *See* docket for *West Virginia, et al. v. EPA*, No. 15-1363 (D.C. Cir. docketed Oct. 23, 2015). For further information on the legal challenge to the CPP, see CRS Report R44480, *Clean Power Plan: Legal Background and Pending Litigation in West Virginia v. EPA*, by Linda Tsang and Alexandra M. Wyatt.

[66] *See generally* Opening Br. of Pet'rs on Core Legal Issues, at 29-61 (filed Feb. 19, 2016). For further information on the legal issues raised in the CPP litigation, see CRS Report R44480, *Clean Power Plan: Legal Background and Pending Litigation in West Virginia v. EPA*, by Linda Tsang and Alexandra M. Wyatt.

[67] See "A Selection of Arguments on the Merits: Section 112 Exclusion" section of CRS Report R44480, *Clean Power Plan: Legal Background and Pending Litigation in West Virginia v. EPA*, by Linda Tsang and Alexandra M. Wyatt for additional information on this issue.

[68] Order at 2, *West Virginia v. EPA*, No. 15-1363 (D.C. Cir. May 16, 2016).

emission sources in areas classified as attainment or unclassifiable as to a NAAQS.[69] For new major emitting facilities and new major modifications, this program requires *new source review*—in particular, preconstruction permits conditioned on installation of the *best available control technology* (BACT).[70]

BACT is determined by the states, with review by EPA, and thus may vary somewhat from state to state. However, BACT must be at least as stringent as the NSPS and Section 112 (hazardous air pollutant) standards for the pollutant,[71] and the PSD new source review program uses the same definitions as to what constitutes a major source, what is a modification, etc., as the NSPS program.[72] Because states can impose more stringent requirements than NSPS, and may need to do so for nonattainment areas to demonstrate attainment, there may be states or areas with less stringent and more stringent emission requirements. However, by imposing NSPS and hazardous air pollutant standards as a floor in all states, this disparity is minimized.

The early D.C. Circuit decision in *Alabama Power Co. v. Costle* remains the fullest judicial exposition of new source review in PSD areas.[73] Among other things, the decision held that new source review is required only for sources to be constructed *in* PSD areas, not in other areas based on a source's projected adverse impacts in a PSD area in another state.[74] *Alabama Power* also held that the PSD sections of the CAA impose no requirement for *post*-construction monitoring of emissions.[75] Finally, the decision approved EPA's use of the "bubble concept" to define which changes in a stationary source constitute a "modification." (See later section on "Bubble Concept.")

The Supreme Court has weighed in on the federal-state relationship in determining BACT. In *Alaska Dep't of Environmental Conservation v. EPA*,[76] the high court said that the CAA authorizes EPA to block construction in a PSD area of a major emitting facility, despite the state's issuance of a PSD permit, when EPA finds that the state's determination of BACT for that facility is inconsistent with the CAA definition of BACT.[77] The state had argued that the Act limits EPA to reviewing whether the state's permit contained a BACT determination.[78]

In 2014, the Supreme Court took up the issue of whether PSD new source review applies in the special case of *greenhouse gas* (GHG) emissions. EPA had argued that once its

---

[69] CAA §§ 165-169; 42 U.S.C. §§ 7475-7479.

[70] CAA § 165(a)(4); 42 U.S.C. § 7475(a)(4).

[71] CAA § 169(3); 42 U.S.C. § 7479(3).

[72] Note that Section 169(2)(C) states that "modification" in the PSD section shall be construed "as defined in" the NSPS section. In *Environmental Defense v. Duke Energy Corp.*, 549 U.S. 561, 576 (2007), however, the Supreme Court held that this common *statutory* definition does not remove EPA discretion to define the term differently in *regulations* under the two sections of the statute, if context so requires.

[73] 636 F.2d 323 (D.C. Cir. 1979).

[74] *Id.* at 364-366.

[75] *Id.* at 373.

[76] 540 U.S. 461 (2004).

[77] *Id.* at 492-493. The CAA definition of BACT is in Section 169(3), 42 U.S.C. § 7479(3).

[78] 540 U.S. at 488.

regulations limiting GHG emissions from new motor vehicles[79] took effect in 2011, it was required to apply PSD new source review to GHG emissions from stationary sources as well. Its argument had some force: the PSD portion of the Act defines the new sources to which it applies as those emitting more than a certain amount of "any *air pollutant*"[80] and requires BACT for "each *air pollutant* subject to regulation under this act."[81] In 2007, the Supreme Court had squarely held that GHGs are indeed "air pollutants" under the CAA.[82] The problem with this argument, however, was administrative unwieldiness. EPA's view concededly meant that tens of thousands of new and modified stationary sources would now require PSD permits, owing to the low statutory emissions thresholds that trigger those permitting requirements and the huge number of sources that emit above-threshold quantities of CO_2, the primary GHG. Moreover, millions of existing stationary sources—office buildings, large apartment buildings, hospitals, etc.— would require Title V operating permits,[83] for the same reason. EPA proposed to take care of this problem with its "Tailoring Rule,"[84] under which EPA raised the emission thresholds that would trigger PSD and Title V permitting, starting with the largest emitters, and then gradually would phase in lower thresholds.

Notwithstanding EPA's syllogistic argument for eventual full coverage of GHGs under PSD and Title V, the Supreme Court took a different tack in *Utility Air Regulatory Group v. EPA.*[85] As the Court explained, just because "air pollutant" as used in the CAA generally covers GHGs does not mean it includes GHGs every place it is used—especially since EPA has historically adopted narrow readings of "air pollutant" in specialized CAA contexts.[86] Moreover, the administrative unwieldiness of demanding PSD and Title V permits for so many sources argued strongly, in the Court's view, against a GHG-inclusive reading of those programs.[87] Nor did the Court allow EPA, through its Tailoring Rule, to phase in the low statutory emission thresholds in an effort to ease the daunting administrative workload, since the Act states the thresholds in absolute numerical terms.[88]

Yet the Court softened its ruling by holding that when new source review, with its BACT requirement, is required because a new or modified source emits a *conventional* pollutant, then such review can be demanded for GHG emissions from that source as well.[89] According to the Solicitor General's numbers submitted in the case, such "anyway"

---

[79] 75 Fed. Reg. 25,323 (May 7, 2010).
[80] CAA § 169(1); 42 U.S.C. § 7479(1) (definition of "major emitting facility"). Emphasis added.
[81] CAA § 165(a)(4); 42 U.S.C. § 7475(a)(4). Emphasis added.
[82] *Massachusetts v. EPA*, 549 U.S. 497 (2007).
[83] CAA § 502; 42 U.S.C. § 7661a.
[84] 75 Fed. Reg. 31,514 (June 3, 2010).
[85] 134 S. Ct. 2427 (2014).
[86] *Id.* at 2439-2442.
[87] *Id.* at 2442-2444. Said the Court: "the PSD program and Title V are designed to apply to, and cannot be extended beyond, a relative handful of large sources capable of shouldering heavy substantive and procedural burdens." *Id.* at 2443.
[88] *Id.* at 2444-2446.
[89] *Id.* at 2444-2449.

sources, so called because they are covered independently of their GHG emissions, account for roughly 83% of American stationary source GHG emissions. Had EPA's Tailoring Rule survived, EPA would have reached only 3% more of those emissions. Thus, on the primary *UARG* holding, EPA was largely victorious, even though its Tailoring Rule was rejected.

"ROUTINE MAINTENANCE"

As mentioned, NSPSs and new source review in PSD areas apply not just to new sources, but also to existing sources that undergo modifications (NSPS) or major modifications (PSD).[90] The CAA defines "modification" as "any" physical or operational change in a stationary source "that increases the emissions of any air pollutant or results in the emission of any air pollutant not previously emitted."[91] A modification is subject to the same requirements as a new source.

The CAA definition of "modification" has been interpreted by EPA and state pollution control agencies. Most important, EPA since 1974 has construed the term not to include "*routine maintenance, repair, and replacement*" (RMRR) at a stationary source[92]—despite the CAA's inclusion of "any" physical or operational change that increases emissions in its definition of modification.[93] Courts have long accepted this agency-created exemption as reasonable.[94] Given the powerful financial incentives for a source to avoid application of stringent NSPSs, utilities have argued for an expansive reading of the RMRR exemption, extending even to major renovations.[95]

The seminal RMRR judicial pronouncement remains *Wisconsin Elec. Power Co. v. Reilly*,[96] referred to as the *WEPCO* decision. *WEPCO*, decided in 1990 by the Seventh Circuit, was the first time a court held that an electric power plant renovation was not RMRR and thus triggered NSPS and PSD new source review.[97] The case established that determining whether the RMRR exemption applies to an existing-source renovation depends on a case-by-case weighing of four factors: the resulting increase in a plant's life expectancy, and the project's cost, nature, and magnitude.[98] These factors are routinely cited by courts decades after the *WEPCO* decision.[99]

---

[90] In the parlance of the CAA, "new source" is defined to include both new and modified existing sources. CAA § 111(a)(2); 42 U.S.C. § 7411(a)(2).

[91] CAA § 111(a)(4), 42 U.S.C. § 7411(a)(4), defines "modification" for purposes of the NSPS section of the CAA. CAA § 169(2)(C), 42 U.S.C. § 7479(2)(C), specifies that that definition applies as well within the PSD portion of the statute.

[92] 40 C.F.R. § 60.14(e) (as to applicability of NSPS); 40 C.F.R. § 52.21(b)(2)(iii)(*a*) (as to applicability of PSD program).

[93] CAA § 111(a)(2); 42 U.S.C. § 7411(a)(2).

[94] *See, e.g.*, Wisconsin Elec. Power Co. v. Reilly, 893 F.2d 901,906 (7th Cir. 1990).

[95] *See, e.g., United States v. Ohio Edison Co.*, 276 F. Supp. 2d 829, 840-849 (S.D. Ohio 2003).

[96] 893 F.2d 901 (7th Cir. 1990).

[97] *Id.* at 912.

[98] *Id.* at 910-911.

[99] *See, e.g., United States v. Duke Energy Corp.*, 5 F. Supp. 3d 771, 775-76 (M.D.N.C. 2014).

The other major RMRR decision is *New York v. EPA*,[100] rejecting a 2003 EPA effort to expand the exception. In its Equipment Replacement Rule of that year, the agency stated that the replacement of a plant's components is categorically within the RMRR exception if the new equipment does not exceed 20% of the replacement value of the process unit and does not change its basic design parameters.[101] The D.C. Circuit found this rule overbroad—inconsistent with the CAA definition of "modification" as including "any" physical change that increases emissions.[102] Phrased another way, the court found the rule too broad to fit within the *de minimis* rationale for the RMRR exception.

In *Environmental Defense v. Duke Energy Corp.*,[103] the Supreme Court took on the question of what constitutes a "modification," a threshold issue for application of the RMRR exception under either NSPS or PSD new source review. The Court held that even though the CAA defines "modification" in its PSD portion to mean the same as the NSPS definition of the term,[104] EPA did not have to define the term the same in each CAA program, owing to the different statutory contexts.[105] EPA, the Court held, can define the definitional phrase "any change in the method of operation of a stationary source"[106] by different measures of the amount of pollutant emitted. To this day EPA defines the emissions increase required by the definition of "modification" differently in the NSPS and PSD contexts: in terms of emissions amount per hour for NSPS,[107] in terms of emissions amount per year for PSD.[108]

BUBBLE CONCEPT

In three different CAA programs—NSPS, PSD, and nonattainment areas—the Act attaches regulatory consequences to modifications at an emissions source that increase the amount of an emitted regulated pollutant.[109] In the CAA's early years, a central question was whether the determination as to whether an emissions increase occurred focused on each individual unit at a plant, or instead on the net aggregate effect of contemporaneous changes within the same source.[110] The latter view was dubbed the *bubble concept* since it

[100] 443 F.3d 880 (D.C. Cir. 2006).

[101] 68 Fed. Reg. 61,248, 61,270 (Oct. 27, 2003). *See also* 70 Fed. Reg. 33,838 (June 10, 2005).

[102] 443 F.3d at 890.

[103] 549 U.S. 561 (2007).

[104] The NSPS definition is at CAA § 169(2)(B); 42 U.S.C. § 7479(2)(B).

[105] 549 U.S. at 574.

[106] *Id.* at 578-579.

[107] 40 C.F.R. § 60.14(b).

[108] 40 CFR §§ 52.21(b)(2)(i), 52.21(b)(23)(i).

[109] As noted in the preceding section of this report, applicability of NSPSs is triggered by, among other things, "modifications" of existing stationary sources. "Modification" is defined for NSPS purposes as a physical or operational change in an existing stationary source that "increases" the amount of emissions therefrom. Both the PSD and nonattainment portions of the CAA are also triggered by modifications of existing stationary sources, and both incorporate the NSPS definition of the term. CAA § 168(2)(C), 42 U.S.C. § 7479(2)(C) (PSD); CAA § 171(4), 42 U.S.C. § 7501(4) (nonattainment).

[110] For a summary of the rulemaking proceedings, see *ASARCO v. EPA*, 578 F.2d 319, 322-325 (D.C. Cir. 1978).

imagines a huge dome—that is, a bubble—placed over a facility, with a single emissions point at the top. Because the amount of emissions at the top of the dome is unaffected by emission increases at individual plant units that are offset by emissions decreases elsewhere in the plant, the bubble image was used to support measuring emission amounts on an aggregate plantwide basis. The bubble concept is preferred by industry because it allows a plant to avoid stringent CAA regulation triggered by emission increases by finding offsetting emission decreases.

By regulation, EPA adopted the bubble concept in each of the three programs listed above, and on each occasion was challenged. In *ASARCO v. EPA*, the D.C. Circuit in 1978 found the CAA barred the bubble concept for determining when a "modification" to an existing stationary source occurred for purposes of triggering application of NSPSs.[111] (As noted earlier, a modification to an existing source triggers NSPSs the same as construction of a new source.) The court saw EPA's bubble regulations, which defined "stationary source" to include any "*combination* of ... facilities,"[112] to be irreconcilable with the CAA definition of "stationary source" as "any building, structure, facility, or installation."[113] Nor was the court moved by EPA's argument that the flexibility afforded by the bubble concept was needed because the cost of bringing modified existing facilities into compliance with NSPSs was, EPA claimed, much greater than the cost of bringing new facilities into compliance.[114]

The following year, the D.C. Circuit further addressed the bubble concept. In *Alabama Power Co. v. Costle*, the same court that found the bubble concept unacceptable for determining whether NSPSs apply held that the bubble concept is required in PSD areas.[115] The alternate view, based on individual units within a plant, was, in the court's view, "unreasonable and contrary to the expressed purposes of the PSD provisions of the Act" and "extremely burdensome."[116] The court distinguished its earlier holding, in *ASARCO*, by pointing out differences in EPA's NSPS and PSD bubble regulations and noting the differing statutory purposes of the two programs.[117]

Finally, in 1984, the Supreme Court found EPA's use of the bubble concept in areas not attaining a NAAQS to be grounded on a permissible reading of the CAA. In *Chevron, U.S.A. v. NRDC*,[118] the Court found the Act's language and legislative history to be sufficiently vague as to use of the bubble concept in nonattainment areas as to warrant deference to EPA's interpretation. The only relevant definition, said the Court, is the general CAA definition of "major stationary source," which equates "major stationary

---

[111] *Id.* at 326.
[112] *Id.* at 324 (emphasis added).
[113] CAA § 111(a)(3); 42 U.S.C. § 7411(a)(3).
[114] 578 F.2d at 328.
[115] 636 F.2d 323, 400-403 (D.C. Cir. 1979).
[116] *Id.* at 401.
[117] *Id.* at 397.
[118] 467 U.S. 837 (1984).

source" with "major emitting facility."[119] The Court found it within "common English usage to take a reference to a major facility or a major source to connote an entire plant as opposed to its constituent parts."[120]

NATIONAL EMISSION STANDARDS FOR HAZARDOUS AIR POLLUTANTS

Loosely speaking, "hazardous air pollutants" under the CAA are pollutants that are more toxic than pollutants addressed by NAAQSs.[121] For this reason, they are regulated pursuant to Section 112 of the Act, which directs EPA to set National Emission Standards for Hazardous Air Pollutants (NESHAPs). Section 112 is among the most litigated sections of the CAA.[122] In the 1970s and 1980s, EPA was slow to implement the section, largely because it found its requirements to be unworkable.[123] By 1990, the agency had set standards for only seven hazardous air pollutants (and not all sources of even these seven), despite informal acknowledgement that hundreds of substances might merit emission controls.[124] EPA was challenged frequently in court for its failure to act.[125]

As enacted in 1970, Section 112 directed EPA to issue NESHAPs that protected public health with an "adequate margin of safety." For several EPA Administrators, under both Democratic and Republican Presidents, this language posed a dilemma.[126] Many of the substances that might merit regulation under Section 112 were possible or probable human carcinogens.[127] As there is no known exposure level at which exposure to a carcinogen is considered safe, this implied that emissions of such substances should be completely eliminated (though the D.C. Circuit rejected this view).[128]

This and other issues raised in litigation during the 1980s led Congress to rewrite completely Section 112 in the 1990 CAA amendments.[129] First, instead of requiring the Administrator to identify hazardous air pollutants, the 1990 amendments listed 187 hazardous air pollutants (HAPs) for EPA to regulate—a list EPA may then revise.[130] Second, instead of initially requiring EPA to protect public health with an adequate margin of safety,

---

[119] CAA § 302(j); 42 U.S.C. § 7602(j).
[120] 467 U.S. at 860.
[121] A more formal definition of hazardous air pollutants is at CAA Section 112(d)(2), 42 U.S.C. § 7412(d)(2).
[122] 42 U.S.C. § 7412.
[123] *See generally* Hearing Before the Subcomm. on Oversight and Investigations, House Comm. on Energy and Commerce, 98th Cong., 2d Sess. 21 (1984) (statement of William D. Ruckelshaus, Administrator, EPA).
[124] *See* S. Rep. No. 101-228, at 3 (1989).
[125] *See, e.g., Sierra Club v. Gorsuch*, 551 F. Supp. 785 (N.D. Cal. 1982) (radionuclides); *New York v. Gorsuch*, 554 F. Supp. 1060 (S.D.N.Y. 1983) (arsenic).
[126] *See* Hearing Before the Subcomm. on Oversight and Investigations, *supra* note 122.
[127] *Id.*
[128] *NRDC v. U.S. EPA*, 824 F.2d 1146 (D.C. Cir. 1987) (en banc).
[129] P.L. 101-549, 104 Stat. 2531 (1990).
[130] CAA § 112(b); 42 U.S.C. §7412(b).

the 1990 amendments require the Administrator, as a first step, to impose *technology-based* emission standards, called "floors," for both new and existing sources. For "major sources," these standards must require use of "maximum achievable control technology" (MACT);[131] for other sources, termed "area sources" by the Act, EPA may elect to impose more lenient standards.[132] In the second, *risk-based* stage of regulation, EPA must review any residual health risks not eliminated by the foregoing standards, and report them to Congress.[133] If Congress does not act, EPA must impose standards providing an ample margin of safety to protect public health, or a stricter standard if needed (considering costs) to prevent adverse environmental effects.[134] These are called "beyond the floor" standards.

Because of ambiguities in the CAA definition of "major source," the D.C. Circuit held in *National Mining Ass'n v. EPA*[135] that EPA may determine if a facility is a major source by aggregating all sources in a contiguous plant site under common control; EPA is not restricted to aggregating only sources within a single source category or under the same Standard Industrial Classification Code.[136] Elsewhere, the court addressed the CAA directive that EPA, in determining what facilities are major sources, calculate whether a facility's "potential to emit" meets the Act's threshold quantities only after "considering controls."[137] The court found that the directive does not allow EPA to limit itself to controls that are federally enforceable.[138]

In *National Lime Ass'n v. EPA*, the D.C. Circuit found that Section 112 requires EPA to create a NESHAP for a hazardous air pollutant in a source category even if no sources in the category use technological controls for that pollutant.[139] Nothing in the statute, the court said, suggests that EPA may set emission levels only for listed hazardous air pollutants that are currently controlled with technology.[140] To the contrary, the D.C. Circuit noted, the CAA requires EPA to set emission standards for each category or subcategory of major sources of hazardous air pollutants listed for regulation.[141]

In *NRDC v. EPA*,[142] the D.C. Circuit wrestled with an ambiguous CAA instruction as to second-stage risk-based standards. The statute says that if the initial, technology-based standard for a source category does not reduce lifetime excess cancer risk to less than 1 in 1 million, EPA "shall promulgate standards ... for such source category."[143] Because this instruction is silent about the stringency of those standards and is "deliberately

---

[131] CAA § 112(d)(2); 42 U.S.C. § 7412(d)(2). "Major source" is defined in CAA Section 112(a)(1).

[132] CAA § 112(d)(5); 42 U.S.C. § 7412(d)(5). "Area source" is defined in CAA Section 112(a)(2) as any stationary source of hazardous air pollutants that is not a "major source."

[133] CAA § 112(f); 42 U.S.C. § 7412(f).

[134] CAA § 112(f)(2)(A); 42 U.S.C. § 7412(f)(2)(A).

[135] 59 F.3d 1351 (D.C. Cir. 1995).

[136] *Id.* at 1359.

[137] CAA § 112(a)(1); 42 U.S.C. § 7412(a)(1).

[138] 59 F.3d at 1364.

[139] 233 F.3d 625, 633-634 (D.C. Cir. 2000).

[140] *Id.* at 633.

[141] *Id.*

[142] 529 F.3d 1077 (D.C. Cir. 2008).

[143] CAA § 112(f)(2)(A); 42 U.S.C. § 7412(f)(2)(A).

ambiguous,"[144] the court upheld EPA's reaffirmation of its existing technology-based standard for the source category in question under which no individual would face an excess lifetime cancer risk of greater than 100 in 1 million. The court noted that 100 in 1 million was EPA's interpretation of "ample margin of safety" before the 1990 amendments, and that those amendments expressly disavow any intent to change the agency's pre-1990 interpretation.[145] The court similarly upheld EPA's consideration of costs in setting the risk-based standards, since the agency had considered costs before 1990.[146]

Although EPA may consider cost when setting second-stage risk-based standards, whether EPA may consider costs when determining if it is "appropriate and necessary" to regulate power plants under Section 112 has been subject to significant litigation. In Section 112, Congress created a special procedure for applying the NESHAP program to power plants.[147] Congress directed EPA to study the public-health hazards power plant HAP emissions pose after taking into account other CAA requirements controlling power plant emissions and allowed EPA to regulate HAP emissions from power plants under Section 112 only if EPA found that "regulation [was] appropriate and necessary" after considering the results of the study.[148]

In 2012,[149] EPA found that it was "appropriate and necessary" to regulate coal- and oil-fired power plants under Section 112 because mercury emissions from power plants (the largest domestic source of mercury emissions) present significant hazards to public health and the environment, and the CAA's other requirements had failed to sufficiently reduce the health risks.[150] In this finding, EPA interpreted the statutory phrase "appropriate and necessary" to preclude it from considering cost when deciding whether to regulate power plants under Section 112.[151] EPA then issued the Mercury and Air Toxics Standard (known as the "MATS rule"), which limits emissions of mercury and other toxics emitted by power plants.[152]

Twenty-three states, along with numerous industry groups, sought review of EPA's MATS rule, arguing, among other things, that EPA acted unlawfully by refusing to

---

[144] 529 F.3d at 1081.

[145] *Id.* at 1082. CAA § 112(f)(2)(B); 42 U.S.C. § 7412(f)(2)(B).

[146] 529 F.3d at 1084.

[147] CAA § 112(n)(1)(A); 42 U.S.C. § 7412(n)(1)(A).

[148] *Id.*

[149] Prior to the 2012 finding, EPA concluded in 2000 that the regulation of power plants was "appropriate and necessary." Regulatory Finding on the Emissions of Hazardous Air Pollutants from Electric Utility Steam Generating Units, 65 Fed. Reg. 79,825, 79,826, 79,830 (Dec. 20, 2000). However, EPA attempted to revoke its "appropriate and necessary" finding in 2004 so that the agency could regulate power plants under Section 111(d). Standards of Performance for New and Existing Stationary Sources: Electric Utility Steam Generating Units, 70 Fed. Reg. 28,606, 28,610, 28,624-32 (May 18, 2005). The rule was challenged, and the D.C. Circuit held that Section 111(d) cannot be used to regulate sources listed under Section 112 and vacated the rule. *See* New Jersey v. EPA, 517 F.3d 574, 580, 583 (D.C. Cir. 2008).

[150] National Emission Standards for Hazardous Air Pollutants from Coal- and Oil-Fired Electric Utility Steam Generating Units, 77 Fed. Reg. 9,304, 9,363 (Feb. 16, 2012).

[151] *Id.* at 9326–27.

[152] National Emission Standards for Hazardous Air Pollutants from Coal- and Oil-Fired Electric Utility Steam Generating Units, 77 Fed. Reg. 9304, 9363 (Feb. 16, 2012).

consider cost in making its "appropriate and necessary" finding.[153] In a per curiam opinion, a divided panel of the D.C. Circuit upheld EPA's decision, holding that EPA's refusal to consider costs was a permissible interpretation of the phrase "appropriate and necessary" when it determined to regulate power plants under Section 112.[154] The court reasoned that Section 112(n)(1) on its face neither requires nor prohibits EPA from considering costs, and that "[t]hroughout section 112, Congress mentioned costs explicitly where it required EPA to consider them."[155] The court made clear that once the "appropriate and necessary" determination is made, EPA must consider costs if it seeks to set emission standards more stringent than those dictated by the statutory MACT standard.[156]

In 2015, the Supreme Court reversed *Michigan v. EPA*.[157] The Court held that EPA's failure to consider costs as a part of its "appropriate and necessary" finding was an unreasonable interpretation of the CAA.[158] Although the Court agreed that EPA's interpretation of the ambiguous term "appropriate and necessary" was entitled to deference under *Chevron U.S.A., Inc. v. NRDC*, the Court concluded that EPA had "strayed far beyond [the bounds of reasonable interpretation] when it read § 7412(n)(1) to mean that it could ignore cost when deciding whether to regulate power plants."[159] The Court explained that

> Read naturally in . . . context, the phrase "appropriate and necessary" requires at least some attention to cost. . . . Agencies have long treated cost as a centrally relevant factor when deciding whether to regulate. . . . [I]t is unreasonable to read an instruction to an administrative agency to determine whether "regulation is appropriate and necessary" as an invitation to ignore cost.[160]

The dissenting Justices agreed that EPA was unreasonable in not taking into account costs in adopting the regulations, but they concluded that EPA had satisfied this requirement when it considered costs when setting the actual emissions standards.[161]

In response to *Michigan v. EPA*, EPA issued a supplemental finding in 2016, concluding that a consideration of cost did not change its determination that regulation of HAP emissions from power plants is "appropriate and necessary" under Section 112.[162] Utilities, industry groups, and 14 states have challenged EPA's finding while several states and local governments, along with two energy companies, have intervened in support of EPA.[163] The petitioners question, among other things, EPA's cost-benefit methodology and whether

---

153 White Stallion Energy Ctr., *LLC v. EPA*, 748 F.3d 1222, 1236 (D.C. Cir. 2014) (per curiam).
154 *Id.* at 1241.
155 *Id.* at 1237.
156 *Id.* at 1240.
157 135 S. Ct. 2699 (2015).
158 *Id.* at 2707.
159 *Id.*
160 *Id.* at 2706.
161 *Id.* at 2716-17 (Kagan, J., dissenting).
162 Supplemental Finding that It is Appropriate and Necessary to Regulate Hazardous Air Pollutants From Coal- and Oil-Fired Electric Utility Steam Generating Units; Final Rule, 81 Fed. Reg. 24,420 (Apr. 25, 2016).
163 *Murray Energy Corp. v. EPA*, No. 16-1127 (D.C. Cir. docketed Apr. 25, 2016).

believes a violation of the statute has occurred: ask the Department of Justice (DOJ) to file a court action for civil penalties: ask the DOJ to do the same for criminal penalties; conduct an EPA adjudication and impose civil penalties; and finally, issue an ACO.[185] All of these options, save ACOs, give the accused party the right to challenge EPA's understanding of the law or the facts in court before penalties can be imposed. By contrast, EPA may seek potentially severe civil penalties or criminal penalties for noncompliance with an ACO even though EPA issues ACOs based on "any information available" and absent adjudication. Such penalties are in addition to penalties for violations of the Act itself. In reviewing an ACO noncompliance case, a court is limited to the question of whether the party complied with the ACO.

In *TVA*, the U.S. Court of Appeals for the Eleventh Circuit struck down this scheme as "repugnant to the Due Process Clause of the Fifth Amendment."[186] Pre-enforcement review by a court, it held, must be made available to the recipient of the order.[187] Nor can EPA "save" the statute by conducting a voluntary adjudication, as before its internal Environmental Appeals Board.[188] In the Circuit's view, that would relegate Article III district courts to insignificant tribunals serving merely as forums for EPA to conduct show-cause hearings simply on whether the party complied with the ACO.[189] Parenthetically, the Supreme Court later rendered an identical ruling—that ACOs are subject to pre-enforcement court review—but in connection with the Clean Water Act and solely on statutory, rather than constitutional, grounds.[190]

In *Dow Chemical Co. v. United States*,[191] the Supreme Court held that CAA Section 114[192] authorizes EPA to use aerial photography in checking on emission sources.[193] Moreover, the Court rejected that the Fourth Amendment required a warrant for such surveillance on the theory that expectations of privacy outside an industrial plant are not akin to those surrounding a residential home.[194]

---

[185] CAA § 113(a); 42 U.S.C. § 7413(a).
[186] 336 F.3d at 1258.
[187] *Id.*
[188] *Id.* at 1259.
[189] *Id.*
[190] *Sackett v. EPA*, 566 U.S. 120 (2012) (holding that Clean Water Act ACOs are "final agency actions" for which pre-enforcement review is available under the Administrative Procedure Act).
[191] 476 U.S. 227 (1986).
[192] 42 U.S.C. § 7414.
[193] 476 U.S. at 239.
[194] *Id.* at 238-239.

In: The Clean Air Act and The Clean Power Plan
Editor: Claire Ferguson
ISBN: 978-1-53614-825-1

Chapter 5

THE CLEAN AIR ACT'S GOOD NEIGHBOR PROVISION: OVERVIEW OF INTERSTATE AIR POLLUTION CONTROL*

Kate C. Shouse

Notwithstanding air quality progress since 1970, challenges remain to reduce pollution in areas exceeding federal standards and to ensure continued compliance elsewhere. The movement of air pollutants across state lines, known as interstate transport, has made it difficult for some downwind states to attain federal ozone and fine particulate matter (PM2.5) standards, partly because states lack authority to limit emissions from other states. The Clean Air Act's "Good Neighbor" provision (Section 110(a)(2)(D)) seeks to address this issue and requires states to prohibit emissions that significantly contribute to another state's air quality problems. It requires each state's implementation plan (SIP)—a collection of air quality regulations and documents—to prohibit emissions that either "significantly contribute" to nonattainment or "interfere with maintenance" of federal air quality standards in another state. The act also authorizes states to petition EPA to issue a finding that emissions from "any major source or group of stationary sources" violate the Good Neighbor provision (Section 126(b)).

EPA and the states have implemented regional programs to address interstate ozone and PM2.5 transport and comply with the Good Neighbor provision. These programs set emission "budgets" for ozone and PM2.5 precursor emissions—specifically, sulfur dioxide

* This is an edited, reformatted and augmented version of a Congressional Research Service report, Publication No. R45299, Updated August 30, 2018.

(SO_2) and nitrogen oxide (NO_x) as PM2.5 precursors and seasonal NO_x emissions as an ozone precursor. The current program—the Cross State Air Pollution Rule (CSAPR)—focuses on limiting interstate transport of power sector SO_2 and NO_x emissions to eastern states.

Power sector emissions in CSAPR states are below emission budgets as a result of regulatory and market factors (see figure). Annual SO_2, annual NO_x, and ozone season NO_x emissions from CSAPR sources decreased 77%, 41%, and 15%, respectively, between 2009 and 2016.

CSAPR Emission Trends: 2009-2016

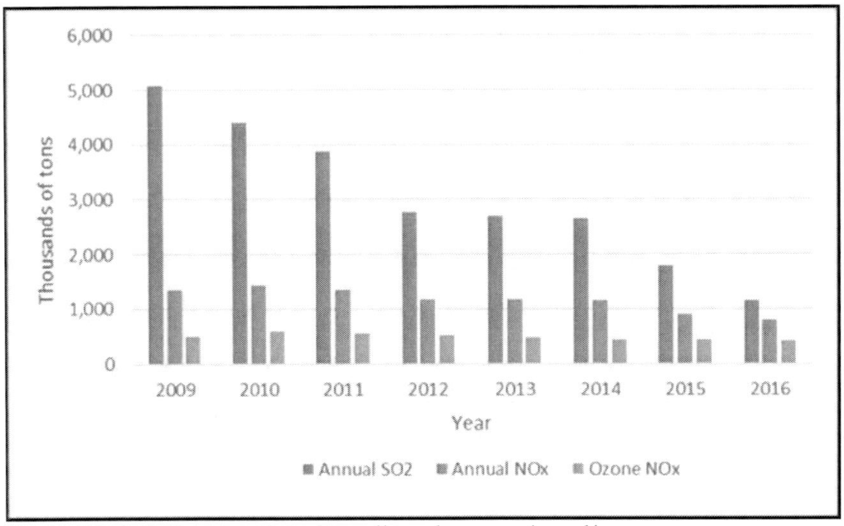

Source: EPA Air Markets Program Data, https://ampd.epa.gov/ampd/.

Notes: The Clean Air Interstate Rule was in effect 2009 through the end of 2014 and was replaced by CSAPR on January 1, 2015.

EPA has concluded that regional SO_2 and NO_x programs have reduced interstate transport of $PM_{2.5}$ and ozone. The Energy Information Administration's national-scale analysis identifies market and regulatory factors contributing to emission reductions. It is unclear whether emissions will remain well below budgets, given recent prices of ozone season NO_x allowances (i.e., authorization for each ton emitted) and the supply of banked allowances for future use in lieu of emission reductions.

Research indicates that ozone transport harms air quality in downwind states. However, stakeholder views vary regarding the extent to which interstate transport impacts air quality. Some note that some coal-fired power plants do not fully use already-installed pollution controls. Several states have sought additional upwind reductions in ozone precursors through Section 126(b) petitions. Others have questioned the feasibility of achieving additional reductions in ozone precursors, raising concerns about emissions from international or natural sources.

EPA recently proposed to determine that CSAPR fully addresses Good Neighbor obligations for the 2008 ozone standard but has not yet made a "Good Neighbor" determination for the more stringent 2015 ozone standard. The agency has therefore not yet determined whether and how it will update the CSAPR budgets with respect to the 2015 ozone standard.

Members of Congress may have an interest in better understanding how EPA and states implement the Clean Air Act's Good Neighbor provision, particularly as EPA continues its assessment of Good Neighbor obligations under the 2015 ozone standard.

The following issues, among others, may inform deliberations about interstate air transport. First, the extent to which existing programs will improve air quality in areas not meeting the 2015 ozone standard is to be determined. CSAPR has not addressed NO_x emissions from nonpower sector sources, such as large industrial boilers. EPA concluded that industrial sources have potential to cost-effectively reduce NO_x emissions but is less certain about the structure of potential NO_x control strategies. Second, some have questioned whether additional regulatory incentives are necessary to fulfill Good Neighbor obligations, particularly given current NO_x allowance prices. These prices are below the marginal abatement cost, which may result in higher emissions. Third, EPA's current air quality initiatives may indirectly affect its Good Neighbor assessments. EPA recently sought comment on potential flexibilities for the development of Good Neighbor SIPs.

INTRODUCTION

The movement of air pollutants across state lines, known as interstate transport, has posed a decades-long challenge to air quality protection. The Clean Air Act (CAA) assigns responsibility to states to limit emissions from sources within their borders as needed to attain federal health-based air quality standards. A state's air quality, however, may be affected by emissions from upwind sources located in a different state. Hence, controlling emissions within the border of a state may not be sufficient to attain the air quality standard. The downwind state lacks authority to limit emissions from the sources in the upwind state(s) but is nonetheless responsible for attaining the federal standards.

Interstate transport has made it difficult for some downwind states to attain federal standards for ozone and fine particulate matter ($PM_{2.5}$). Both of these pollutants are formed by precursor emissions that can travel long distances. Specifically, sulfur dioxide (SO_2) and nitrogen oxide (NO_x) contribute to the formation of $PM_{2.5}$ in the air.[1] NO_x and volatile

[1] $PM_{2.5}$ is also directly emitted by sources (e.g., construction sites, unpaved roads, smokestacks, or fires). See U.S. Environmental Protection Agency (EPA), *Particulate Matter (PM) Basics*, https://www.epa.gov/pm-pollution/particulate-matter-pm-basics#main-content. For detailed information about PM and the formation of PM, see National Research Council, *Global Sources of Local Pollution: An Assessment of Long-Range Transport of Key Air Pollutants to and from the United States*, 2010, pp. 67-76.

organic compounds (VOCs) react in sunlight to form ground-level ozone, the main component of smog.[2] Studies have shown that these precursor emissions, as well as ozone and $PM_{2.5}$, are regional pollutants, meaning that they can travel hundreds of miles through the atmosphere.[3] For example, Bergin et al.'s study of the eastern United States concluded that regional transport affected air quality in most eastern states. They attributed an average of 77% of each state's ozone and $PM_{2.5}$ concentrations to emissions from upwind states.[4]

These regional emissions are associated with health impacts and are therefore of concern. For example, research shows that ground-level ozone is associated with aggravated asthma, chronic bronchitis, heart attacks, and premature death.[5] Studies have also linked exposure to particulate matter to respiratory illnesses, such as aggravated asthma, as well as heart attacks and premature death.[6]

The CAA's "Good Neighbor" provision recognizes such interstate issues and requires states to prohibit emissions that significantly contribute to air quality problems in another state (Section 110(a)(2)(D)). It requires each state's implementation plan—a collection of air quality regulations and documents—to include adequate provisions to prohibit emissions that either "contribute significantly" to nonattainment or "interfere with maintenance" of federal air quality standards in another state.[7]

Since the 1990s, the U.S. Environmental Protection Agency (EPA) and the states have implemented various regional programs to address interstate air transport. Many of these programs have since concluded. The current program—the Cross State Air Pollution Rule (CSAPR, pronounced "Casper")—is an emissions trading program for 28 states in the eastern part of the United States. EPA established CSAPR to limit interstate transport of power sector SO_2 and NO_x emissions and help states comply with the 1997 and 2006 $PM_{2.5}$ standards as well as the 1997 and 2008 ozone standards.

EPA has attributed emission reductions to CSAPR and the agency's other emissions trading programs, such as the Acid Rain Program: annual SO_2 emissions from power plants

---

[2] EPA, *Basic Information about Ozone*, https://www.epa.gov/ozone-pollution/basic-information-about-ozone.

[3] EPA, *Interstate Air Pollution Transport*, https://www.epa.gov/airmarkets/interstate-air-pollution-transport. See also EPA, "Cross-State Air Pollution Rule Update for the 2008 Ozone NAAQS," 81 *Federal Register* 74514, October 26, 2016; and EPA, *Fact Sheet. The Cross-State Air Pollution Rule: Reducing the Interstate Transport of Fine Particulate Matter and Ozone*, July 2011, p. 1, https://www.epa.gov/sites/production/files/2016-09/documents/csaprfactsheet.pdf.

[4] Michelle S. Bergin et al., "Regional Air Quality: Local and Interstate Impacts of NO_x and SO_2 Emissions on Ozone and Fine Particulate Matter in the Eastern United States," *Environmental Science & Technology*, vol. 41, no. 13 (2007), pp. 4677-4689. In addition, EPA summarizes studies about regional transport. See EPA, "Cross-State Air Pollution Rule Update for the 2008 Ozone NAAQS," 81 *Federal Register* 74514.

[5] In addition, ground-level ozone is associated with environmental effects, such as negative impacts on forests and crop yields. For information about the health and environmental effects of ozone, see EPA, *Basic Information about Ozone*; and EPA, *Integrated Science Assessment (ISA) of Ozone and Related Photochemical Oxidants*, February 2013, https://cfpub.epa.gov/ncea/isa/recordisplay.cfm?deid=247492.

[6] EPA, *Health and Environmental Effects of Particulate Matter (PM)*, https://www.epa.gov/pm-pollution/health-and-environmental-effects-particulate-matter-pm. See also EPA, *Integrated Science Assessment for Particulate Matter*, December 2009, https://cfpub.epa.gov/ncea/isa/ recordisplay.cfm?deid=216546.

[7] CAA §110(a)(2); 42 U.S.C. §7410(a)(2).

participating in CSAPR were 1.2 million tons in 2016, an 87% reduction from 2005 levels.[8] CSAPR power plants also emitted 420,000 tons of NO_x in the 2016 ozone season, roughly an 80% reduction from the 1990 ozone season NO_x emissions.[9]

Emissions reduction progress notwithstanding, some areas of the country do not meet federal air quality standards for pollutants like ozone and particulate matter. In 2018, EPA designated 52 areas with approximately 200 counties or partial counties as "nonattainment" with respect to the 2015 ozone standard.[10]

Members of Congress representing both downwind and upwind states may have an interest in how EPA and states implement the CAA's Good Neighbor provision, particularly as states begin to develop plans for nonattainment areas to come into compliance with the 2015 ozone standards. Some downwind states with nonattainment areas have attributed their ozone violations—at least in part—to emission sources from upwind states.[11] Downwind states have also expressed concerns that transported air pollution contributes to harmful human health impacts and adversely affects economic growth.[12] For example, a Maryland state agency reported that transport of emissions from upwind states has required Maryland's sources to compensate with "deeper instate emissions reductions," thereby adding economic costs to the state's business community.[13] Upwind states have disagreed with the approach used by EPA to determine whether emissions from upwind sources

[8] EPA reports emissions reductions achieved under several cap-and-trade programs designed to reduce SO_2 and NO_x from power plants, including the Acid Rain Program, the Clean Air Interstate Rule (CAIR), and CSAPR (which replaced CAIR in 2015). EPA reports that most of the SO_2 and NO_x emission reductions since 2005 occurred in response to CAIR. See EPA, *2016 Program Progress*, "Emission Reductions," https://www3.epa.gov/ airmarkets/ progress/reports/pdfs/2016_full_report.pdf, and "Emission Reductions: SO_2 Emission Trends," https://www3.epa.gov/ airmarkets/progress/reports/pdfs/2016_full_report.pdf.

[9] EPA, *2016 Program Progress*, "Emission Reductions: Ozone Season NO_x Emission Trends." https://www3 .epa.gov/ airmarkets/progress/reports/pdfs/2016_full_report.pdf.

[10] On April 30, 2018, EPA designated 51 areas as nonattainment with the 2015 ozone standard. EPA designated one more area as nonattainment on July 17, 2018. For the designations of 51 nonattainment areas, see EPA, "Additional Air Quality Designations for the 2015 Ozone National Ambient Air Quality Standards," 83 *Federal Register* 25776, June 4, 2018. For the July 2018 designation, see EPA, "Additional Air Quality Designations for the 2015 Ozone National Ambient Air Quality Standards—San Antonio, Texas Area," 83 *Federal Register* 35136, July 25, 2018. For more information about the 2015 ozone standard and EPA's recent designations, see CRS Report R43092, *Implementing EPA's 2015 Ozone Air Quality Standards*, by James E. McCarthy and Kate C. Shouse.

[11] In their recommendations to EPA regarding designation of nonattainment areas, Delaware, New Jersey, and Wisconsin attributed ozone violations in their jurisdiction to emissions transported from other states. See state recommendation letters to EPA, https://www. epa.gov/ozone-designations/2015-ozone-standards-state-recommendations-epa-responses-and-technical-support.

[12] For example, see U.S. Senator Richard Blumenthal et al., letter to Honorable Scott Pruitt, Administrator, EPA, February 23, 2018, https://www.blumenthal.senate.gov/imo/ media/doc/2018.02.23%20Letter%20to%20 Pruitt%20re%20CT%20Clean%20Air%20Act%20Petition.pdf; and U.S. Congress, Senate Environment and Public Works Committee, Subcommittee on Clean Air and Nuclear Safety, *Cooperative Federalism Under the Clean Air Act: State Perspectives*, 115th Cong., 2nd sess., April 10, 2018. See testimony of Shawn Garvin, Department of Natural Resources and Environmental Control, p. 3, https://www.epw.senate.gov/ public/cache/files/7/9/79fab6e4ae5d4e6faf30d4cd30d7b3d7/18D52D3F4801CCF26167EC7F17CD4C16.garvi n-testimony04.10.2018.pdf.

[13] Maryland Department of Environment, "Governor Larry Hogan Announces State Lawsuit Against EPA," press release, September 27, 2017, http://news.maryland.gov/mde/ 2017/09/27/governor-larry-hogan-announces-statelawsuit-against-epa/.

contribute to downwind air quality problems. For instance, Ohio's state environmental agency described EPA's transport approach as "deeply flawed," concluding that it would place "an unfair amount of responsibility" on upwind power plants to reduce emissions.[14]

To assist Members and staff in understanding interstate transport issues, this chapter presents background information about the CAA's interstate transport provision, provides a brief history of regional programs leading up to CSAPR, discusses key aspects of the CSAPR program and program results, summarizes the status of Good Neighbor determinations with respect to ozone standards, and concludes with issues for congressional consideration.

BACKGROUND

The CAA requires EPA to establish national standards for air pollutants that meet the criteria in Section 108(a)(1). These pollutants—the "criteria pollutants"—are those that EPA has determined "may reasonably be anticipated to endanger public health or welfare" and whose presence in "ambient air results from numerous or diverse mobile or stationary sources."[15] EPA must design two types of National Ambient Air Quality Standards (NAAQS) for the criteria pollutants. Primary NAAQS must protect public health with an "adequate margin of safety," and secondary NAAQS must "protect public welfare from any known or anticipated adverse effects."[16] The NAAQS are concentration standards measured in parts per million (ppm) by volume, parts per billion (ppb) by volume, and micrograms per cubic meter of air ($\mu g/m^3$).[17] The NAAQS do not set direct limits on emissions but rather define what EPA considers to be clean air for the pollutant in question.[18]

Section 109(d) of the act requires periodic NAAQS reviews. Every five years, EPA must review the NAAQS and the science upon which the NAAQS are based and then revise the NAAQS if necessary. This multi-step process is rarely completed within the five-year review cycle and is often the subject of litigation that results in court-ordered deadlines for

---

[14] Craig Butler, Director, Ohio Environmental Protection Agency, letter to Honorable Gina McCarthy, Administrator, EPA, February 1, 2016, p. 1. See EPA-HQ-OAR-2015-0500-0283 at http://www.regulations.gov.

[15] CAA §108(a)(1), codified at 42 U.S.C. §7408(a)(1). EPA determined that the following pollutants meet the Section 108 criteria: (1) carbon monoxide, (2) lead, (3) nitrogen dioxide, (4) ozone, (5) particulate matter, and (6) sulfur dioxide. For more information about the act's requirements, see CRS Report RL30853, *Clean Air Act: A Summary of the Act and Its Major Requirements*, by James E. McCarthy.

[16] CAA §109(b), codified at 42 U.S.C. §7409(b).

[17] EPA, "NAAQS Table," https://www.epa.gov/criteria-air-pollutants/naaqs-table.

[18] CRS Report R43092, *Implementing EPA's 2015 Ozone Air Quality Standards*, by James E. McCarthy and Kate C. Shouse.

completion of NAAQS reviews.[19] Since January 1997, EPA has completed at least one review for each of the six criteria pollutants (carbon monoxide, lead, nitrogen dioxide, ozone, particulate matter, and sulfur dioxide), with standards being made more stringent for five of the six.[20] Most of the revisions finalized in this time period were for the ozone and particulate matter standards.[21]

The CAA assigns responsibility to states to establish procedures to attain and maintain the NAAQS within their borders. In particular, the act requires each state to submit a new or revised state implementation plan (SIP) to EPA within three years of a NAAQS promulgation or revision.[22] This SIP submission, also known as an "infrastructure SIP," outlines how the state will implement, maintain, and enforce the NAAQS[23] The infrastructure SIP allows EPA to "review the basic structural requirements of [a state's] air quality management program in light of each new or revised NAAQS."[24] Examples of the basic structural requirements include enforceable emission limits, an air monitoring program, an enforcement program, air quality modeling capabilities, and "adequate personnel, resources, and legal authority."[25]

The state's SIP must also address its interstate transport obligations under the CCAA. EPA refers to this section of the SIP submission as the "Good Neighbor SIP." The Good Neighbor SIP must prohibit "certain emissions of air pollutants because of the impact they would have on air quality in other states."[26] Specifically, the state's Good Neighbor SIP must prohibit sources in that state from "emitting any air pollutant in amounts which will ... contribute significantly to nonattainment in, or interfere with maintenance" of a NAAQS in another state.[27]

---

[19] On May 9, 2018, the EPA Administrator announced five principles—one of which was meeting statutory deadlines—for EPA to follow in future NAAQS reviews. See E. Scott Pruitt, EPA Administrator, memorandum to Assistant Administrators, May 9, 2018, https://www.epa.gov/criteria-air-pollutants/back-basics-process-reviewingnational-ambient-air-quality-standards.

[20] EPA has not revised the carbon monoxide (CO) NAAQS since 1985, when it revoked the secondary CO NAAQS. See EPA, *Table of Historical Carbon Monoxide (CO) National Ambient Air Quality Standards (NAAQS)*, https://www.epa.gov/co-pollution/table-historical-carbon-monoxide-co-national-ambient-air-quality-standards-naaqs. For a summary of current NAAQS for all six criteria pollutants and links to the history of standards for each of these pollutants, see EPA, "NAAQS Table."

[21] For a summary of the most recent ozone NAAQS revision, see CRS Report R43092, *Implementing EPA's 2015 Ozone Air Quality Standards*, by James E. McCarthy and Kate C. Shouse. For a summary of the most recent particulate matter NAAQS revision, see CRS Report R42934, *Air Quality: EPA's 2013 Changes to the Particulate Matter (PM) Standard*, by Robert Esworthy.

[22] CAA §110(a)(1); 42 U.S.C. §7410(a)(1).

[23] EPA, *Guidance on Infrastructure State Implementation Plan (SIP) Elements under the Clean Air Act Sections 110(a)(1) and 110(a)(2)*, September 2013, p. 1, https://www. epa.gov/sites/production/files/2015-12/documents/guidance_on_infrastructuresip_elements _multipollutant_final_sept_2013.pdf.

[24] EPA, *Guidance on Infrastructure State Implementation Plan*, p. 2.

[25] EPA, *Guidance on Infrastructure State Implementation Plan*, p. 2.

[26] The Good Neighbor SIP is part of the infrastructure SIP. States with nonattainment areas must also submit "nonattainment SIPs" that address more detailed statutory requirements for coming into attainment. See EPA, "Federal Implementation Plans: Interstate Transport of Fine Particulate Matter and Ozone and Correction of SIP Approvals," 76 *Federal Register* 48216, August 8, 2011. See also EPA, "Notice of Availability of the Environmental Protection Agency's Preliminary Interstate Ozone Transport Modeling Data for the 2015 Ozone National Ambient Air Quality Standard (NAAQS)," 82 *Federal Register* 1734, January 6, 2017.

[27] CAA §110(a)(2)(D)(i)(I); 42 U.S.C. §7410(a)(2)(D)(i)(I).

EPA reviews SIPs to ensure they meet statutory requirements. The agency also has authority to require states to revise their SIPs. Furthermore, the act requires EPA, under certain conditions, to impose sanctions and to issue a Federal Implementation Plan (FIP) if a state fails or declines to submit or implement an adequate SIP.[28]

Recognizing ongoing challenges with ozone transport, the 1990 CAA Amendments established regional planning provisions specific to ozone. For example, CAA Section 184 created a multistate ozone transport region, known as the Ozone Transport Region (OTR), and established the northeast Ozone Transport Commission (OTC) to advise EPA about ozone controls in the OTR.[29] The OTR is comprised of 12 Northeastern and Mid-Atlantic states (Connecticut, Delaware, Maine, Maryland, Massachusetts, New Hampshire, New Jersey, New York, Pennsylvania, Rhode Island, Vermont, certain counties in Northern Virginia) and Washington, DC.[30] The CAA required states in the OTR to impose controls on sources in all specified areas, regardless of attainment status. Such controls included enhanced vehicle inspection and maintenance programs and reasonably available control technology for sources of VOCs.[31]

In addition, CAA Section 176A allows EPA to establish transport regions to address regional pollution problems contributing to violations of a primary NAAQS. The agency must establish a commission, comprised of EPA and state officials, for each transport region that makes recommendations to EPA on appropriate mitigation strategies.

IMPLEMENTING THE GOOD NEIGHBOR PROVISION

The CAA provides two independent statutory authorities to facilitate compliance with the Good Neighbor provision: (1) the SIPs process under Section 110 and (2) a petition process under Section 126(b). While these authorities are separate, they each address the same objective—that is, the Good Neighbor provision in Section 110(a)(2)(D)(i).[32] The remainder of this section describes how these authorities may be used to enforce the Good Neighbor provision.

---

[28] Regarding EPA's authority to issue FIPs, see CAA Section 110(c)(1); 42 U.S.C. Section 7410(c)(1). Regarding the sanctions EPA may impose, see CAA Section 110(m); 42 U.S.C. Section 7410(m).

[29] CAA §184; 42 U.S.C. §7511(c). See also Philip E. Karmel, "Interstate Transport and Regional Approaches to Regulating Air Pollution," in *The Clean Air Act Handbook*, ed. Julie R. Domike and Alec C. Zacaroli, 4th ed. (Chicago: American Bar Association, 2016), pp. 115-136.

[30] CAA §184(a); 42 U.S.C. §7511(c).

[31] CAA §184; 42 U.S.C. §7511(c). For further analysis of this provision, see also Philip E. Karmel, "Interstate Transport and Regional Approaches to Regulating Air Pollution," in *The Clean Air Act Handbook*, ed. Julie R. Domike and Alec C. Zacaroli, 4th ed. (Chicago: American Bar Association, 2016), pp. 115-136.

[32] EPA, "Response to Clean Air Act Section 126(b) Petitions from Delaware and Maryland," 83 *Federal Register* 26675, June 8, 2018.

SIPs Process

As previously noted, a state's SIP must prohibit sources in that state from "emitting any air pollutant in amounts which will ... contribute significantly to nonattainment in, or interfere with maintenance" of a NAAQS in another state.[33] If EPA finds an existing SIP inadequate, it must require the state to revise the SIP.[34] This procedure is known as a "SIP call" and it can be issued to multiple states at the same time. Specifically, EPA must issue a SIP call whenever the agency determines that the SIP is "substantially inadequate to attain or maintain" a particular NAAQS, to ensure that the state's sources do not contribute significantly to a downwind state's nonattainment, or if it is otherwise inadequate to meet any CAA requirement.[35] EPA can also issue a SIP call if states do not meet the CAA Section 184 requirements of the OTR.

Section 126(b) Petitions

Under CAA Section 126(b), any state or political subdivision can petition EPA to issue a "finding that any major source or group of stationary sources emits or would emit any air pollutant in violation" of the Good Neighbor provision.[36] Section 126(b) requires EPA to make a decision within 60 days. If EPA grants the petition, the sources identified in the petition must cease operations within three months unless they comply with emission controls and compliance schedules set by EPA.

While Section 126(b) and a SIP call each enforce the Good Neighbor provision, they differ in their implementation.[37] First, a state or political subdivision must initiate the 126(b) petition, whereas EPA initiates the SIP call. Second, unlike a SIP call, the 126(b) petition is limited to a "major source or group of stationary sources" and cannot be used to address minor or mobile sources.[38] Third, EPA may directly regulate upwind sources when it grants a 126(b) petition, whereas a SIP call results in direct EPA regulation only if EPA issues a FIP in response to a state's failure to respond adequately to the SIP call.

---

[33] CAA §110(a)(2)(D)(i)(I); 42 U.S.C. §7410(a)(2)(D)(i)(I).

[34] CAA §110(k)(5); 42 U.S.C. §7410(k)(5).

[35] CAA §110(k)(5); 42 U.S.C. §7410(k)(5).

[36] CAA §126(b); 42 U.S.C. §7426(b). Section 302(j) defines *major stationary source* as "any stationary facility or source of air pollutants which directly emits, or has the potential to emit, one hundred tons per year or more of any air pollutant (including any major emitting facility or source of fugitive emissions of any such pollutant, as determined by rule by the Administrator)" (CAA §302(j); 42 U.S.C. §7602(j)). Section 126(b) cannot be used to address emissions from mobile sources, such as cars and trucks. Another section of the CAA—Title II (42 U.S.C. §§7521-7590)— contains most of the act's provisions for mobile sources.

[37] CAA §126(b); 42 U.S.C. §7426(b). For additional discussion of the differences between 126(b) petitions and a SIP call, see Philip E. Karmel, "Interstate Transport and Regional Approaches to Regulating Air Pollution," in *The Clean Air Act Handbook*, pp. 133-134.

[38] Karmel, "Interstate Transport," in *The Clean Air Act Handbook*, p. 133.

EPA's review of 126(b) petitions has sometimes coincided with the agency's SIP call process. For example, in 1998, EPA coordinated its review of eight 126(b) petitions when it promulgated a SIP call. EPA acknowledged the distinction between the CAA authorities for the 126(b) petition process and the SIP call but coordinated the two actions because they were both designed to reduce ozone transport in the eastern United States.[39]

States have also submitted 126(b) petitions ahead of the deadlines for Good Neighbor SIPs.[40] For example, in 2011, EPA granted a 126(b) petition from New Jersey, finding that a coal-fired generating station in Pennsylvania contributed significantly to nonattainment with the SO_2 NAAQS in New Jersey.[41] Some considered EPA's approval of this petition to reflect a more expansive interpretation of Section 126 in which 126(b) petitions are not necessarily limited to the time frame of Good Neighbor SIP updates.[42] Whereas EPA had previously considered 126(b) petitions several years after revising a NAAQS—and after making attainment and nonattainment designations for revised standards—EPA approved New Jersey's 126(b) petition before Pennsylvania was required to complete its Good Neighbor SIP for the 2010 revision to the SO_2 NAAQS.[43] EPA promulgated an emissions limit for the generating station that would reduce its SO_2 emissions by 81% and set a compliance deadline of three years.[44] In 2013, the U.S. Court of Appeals for the Third Circuit upheld EPA's interpretation of Section 126, concluding that the CAA allows EPA to make a Section 126 finding independently of the Section 110 SIP process.[45]

States have continued to submit Section 126(b) petitions related to ozone interstate transport. For example, Connecticut, Delaware, Maryland, and New York have submitted 126(b) petitions related to compliance with the 2008 and/or 2015 ozone NAAQS.[46] As of July 2018, EPA has denied the petition from Connecticut and has proposed to deny petitions from Delaware and Maryland.[47] Among the various reasons for denying Connecticut's petition, EPA found that the petition did not reflect current operations at the

---

[39] EPA, "Finding of Significant Contribution and Rulemaking for Certain States in the Ozone Transport Assessment Group Region for Purposes of Reducing Regional Transport of Ozone," 63 *Federal Register* 57361-57362, October 27, 1998.

[40] As previously noted, the Good Neighbor SIP is part of the infrastructure SIP. States with nonattainment areas must also submit "nonattainment SIPs" that address more detailed statutory requirements for coming into attainment. See EPA, "Federal Implementation Plans," 76 *Federal Register* 48216.

[41] EPA, "Final Response to Petition from New Jersey Regarding SO_2 Emissions from the Portland Generating Station," 76 *Federal Register* 69052, November 7, 2011.

[42] Karmel, "Interstate Transport," in *The Clean Air Act Handbook*, p. 134.

[43] For EPA's discussion about the timing of 126(b) petitions and Good Neighbor SIPs, see EPA, "Final Response to Petition from New Jersey Regarding SO_2 Emissions from the Portland Generating Station," 76 *Federal Register* 69053 69056, November 7, 2011. For the 2010 SO_2 NAAQS revision, see EPA, "Primary National Ambient Air Quality Standard for Sulfur Dioxide; Final Rule," 75 *Federal Register* 35520, June 22, 2010.

[44] EPA, "Final Response to Petition from New Jersey Regarding SO_2 Emissions from the Portland Generating Station," 76 *Federal Register* 69066-69068, November 7, 2011.

[45] *GenOn REMA, LLC v. EPA*, 722 F.3d 513, 521-22 (3d Cir. 2013).

[46] EPA, *Ozone National Ambient Air Quality Standards (NAAQS) Section 126 Petitions*, June 2018, https://www.epa.gov/ozone-pollution/ozone-national-ambient-air-quality-standards-naaqs-section-126-petitions.

[47] For the final decision on Connecticut's petition, see EPA, "Response to June 1, 2016 Clean Air Act Section 126(b) Petition from Connecticut," 83 *Federal Register* 16064, April 13, 2018. For the proposed denial of Delaware and Maryland's petitions, see EPA, "Response to Clean Air Act Section 126(b) Petitions from Delaware and Maryland," 83 *Federal Register* 26675, June 8, 2018.

named source—a power plant located in Pennsylvania.[48] In particular, EPA stated that the air quality modeling in Connecticut's petition was based on 2011 emissions data and therefore did not account for subsequent NO_x reductions, noting that the named source "primarily burned natural gas with a low NO_x emission rate in the 2017 ozone season."[49] In addition, EPA conducted its own analysis using the agency's current multi-step framework for determining what constitutes a significant contribution.[50] The agency's analysis did not identify additional "highly cost-effective controls available at the source and thus no basis to determine that [the named source] emits or would emit in violation of the good neighbor provision with respect to the 2008 ozone NAAQS."[51] While EPA "expects the facility to continue operating primarily by burning natural gas in future ozone seasons,"[52] others have expressed concern that there is no "enforceable requirement prohibiting" the named source from switching back to coal.[53]

Similarly, in June 2018, EPA proposed to deny petitions from Delaware and Maryland, in part because EPA found "several elements of the states' analyses ... insufficient to support the states' conclusions."[54] For example, EPA said that Delaware's petitions did "not provide any analysis indicating that Delaware may be violating or have difficulty maintaining the 2008 or 2015 ozone NAAQS in a future year associated with the relevant attainment dates."[55] EPA also noted that Delaware used 2011 emissions data, which EPA characterized as "generally higher than, and therefore not representative of, current and

[48] EPA also concluded that Connecticut's petition did not provide enough information regarding the source's impact on air quality in Connecticut on high ozone days. For example, EPA stated that the petition did not demonstrate whether the days in which emissions from the named source had the largest impact on Connecticut's air quality monitors coincided with the highest modeled ozone days at the Connecticut monitors. EPA, "Response to June 1, 2016 Clean Air Act Section 126(b) Petition from Connecticut," 83 *Federal Register* 16064, 16071-16072, April 13, 2018. See also Connecticut Department of Energy and Environmental Protection, *Petition to the U.S. EPA Pursuant to Section 126 of the Clean Air Act for Abatement of Emissions from the Brunner Island Coal Fired Generating Units in Pennsylvania, as Such Generating Units Significantly Contribute to Nonattainment of, and Interfere with Maintenance of, the 2008 Ozone NAAQS in the State of Connecticut*, June 1, 2016.

[49] EPA, "Response to June 1, 2016 Clean Air Act Section 126(b) Petition from Connecticut," 83 *Federal Register* 16072, April 13, 2018.

[50] Regulatory actions and litigation have led to EPA's establishment of the current framework that it uses to assess Good Neighbor obligations. The framework establishes a screening threshold and accounts for both cost and air quality improvements. See "Framework to Assess Good Neighbor Provision" in this report for complete discussion.

[51] EPA, "Response to June 1, 2016 Clean Air Act Section 126(b) Petition from Connecticut," 83 *Federal Register* 16065.

[52] EPA, "Response to June 1, 2016 Clean Air Act Section 126(b) Petition from Connecticut," 83 *Federal Register* 16072.

[53] Arthur N. Marin, Executive Director, Northeast States for Coordinated Air Use Management, letter to William Wehrum, Assistant Administrator, Office of Air and Radiation, EPA, March 26, 2018, http://www.nescaum.org/documents/nescaum-comments-ct-126-petition-20180326.pdf/.

[54] EPA also proposed to find that, based on its own analysis, there are "no additional highly cost-effective emissions reductions available at the sources, and, thus, that none of the named sources currently emit or would emit in violation of the good neighbor provision with respect to the relevant ozone standard." EPA, "Response to Clean Air Act Section 126(b) Petitions from Delaware and Maryland," 83 *Federal Register* 26677, June 8, 2018.

[55] For complete discussion, see EPA, "Response to Clean Air Act Section 126(b) Petitions from Delaware and Maryland," 83 *Federal Register* 26676, June 8, 2018; and Delaware's four 126(b) petitions (Brunner Island, Conemaugh, Harrison, Homer City), available at Delaware Department of Natural Resources and Environmental Control, *Delaware's 126(b) Petitions for Clean Air*, http://www.dnrec.delaware.gov/Air/Pages/126-Petitions.aspx.

future projected emissions levels at these [named sources] and in the rest of the region."[56] Delaware has disagreed with EPA's proposed denial on various grounds. Among other things, Delaware stated that EPA has not "shown valid modeling or justification that Delaware will attain the 2015 ozone standard by its 2021 Marginal nonattainment deadline."[57] In particular, EPA's projections analyzed the year 2023, which is the attainment deadline for areas designated as moderate nonattainment with respect to the 2015 ozone standard.[58]

Finally, EPA proposed to deny Maryland's petition, in part because the agency disagreed with Maryland that NO_x limits for 36 named sources should be based on the respective units' lowest observed emissions rates.[59] Specifically, Maryland's petition concluded that the 36 named sources were operating pollution controls "sub-optimally based on a comparison of their lowest observed NO_x emissions rates between 2005 and 2008, which Maryland describes as the 'best' observed emissions rates, to emissions rates from the 2015 and 2016 ozone seasons."[60] EPA disagreed that the lowest historical NO_x emissions rate is representative of "ongoing achievable NO_x rates" in part because over time, some NO_x controls (e.g., selective catalytic reduction [SCR] systems) "may have some broken-in components and routine maintenance schedules entailing replacement of individual components." EPA stated that in a 2016 rulemaking addressing regional ozone transport, the agency determined that the "third lowest fleetwide average coal-fired [power plant] NO_x rate" for power plants using SCR to be "most representative of ongoing, achievable emission rates."[61] Maryland has disagreed with EPA's proposed denial, noting that it will "testify in opposition to the proposal and use all available tools, including litigation."[62]

---

56 EPA, "Response to Clean Air Act Section 126(b) Petitions from Delaware and Maryland," 83 *Federal Register* 26676, June 8, 2018. Delaware has responded that EPA's 2017 data show two of the named sources operating with emissions rates similar to those in 2011, see Delaware Department of Natural Resources and Environmental Control, *Testimony on EPA's Proposed Denial of Delaware's 126(b) Petitions*, June 22, 2018, p. 2, http://www.dnrec.delaware.gov/Air/Documents/06-22-Dave-Fees-testimony-at-EPA-public -hearing.pdf.

57 Delaware Department of Natural Resources and Environmental Control, *Testimony on EPA's Proposed Denial*, p. 3.

58 EPA, "Response to Clean Air Act Section 126(b) Petitions from Delaware and Maryland," 83 *Federal Register* 26678. See also EPA, *Ozone NAAQS Timelines*, https://www.epa.gov/ ozone-pollution/ozone-naaqs-timelines.

59 For complete discussion, see EPA, "Response to Clean Air Act Section 126(b) Petitions from Delaware and Maryland," 83 *Federal Register* 26677, and Maryland Department of Environment, *Petition to the U.S. EPA Pursuant to Section 126 of the Clean Air Act for Abatement of Emissions from 36 Coal-Fired Electric Generating Units at 19 Plants in Five States that Significantly Contribute to Nonattainment of, and Interfere with Maintenance of, the 2008 Ozone NAAQS in the State of Maryland*, November 16, 2016, http://news.maryland.gov/mde/wp-content/uploads/sites/6/2016/11/MD_126_Petition_ Final_111616.pdf.

60 EPA, "Response to Clean Air Act Section 126(b) Petitions from Delaware and Maryland," 83 *Federal Register* 26677.

61 EPA, "Response to Clean Air Act Section 126(b) Petitions from Delaware and Maryland," 83 *Federal Register* 26677.

62 Maryland Department of Environment, "Maryland Environment Secretary Grumbles Statement on EPA's Proposed Denial of the Maryland Department of the Environment's Clean Air Act Petition," press release, June 1, 2018, http://news.maryland.gov/mde/2018/06/01/maryland-environment-secretary-grumbles-statement-on-epas-proposeddenial-of-the-maryland-department-of-the-environments-clean-air-act-petition/.

Interpreting "Significant Contribution"

Enforcement of the CAA's interstate transport provisions hinges on a key test in Section 110(a)(2)(D)(i)—whether one state "significantly contributes" to a violation of the NAAQS in another state. The CAA does not, however, define what constitutes a significant contribution. Instead, this phrase has been interpreted through EPA rulemakings addressing interstate air pollution. The agency's interpretation has been contentious at times, given that it "inherently involves a decision on how much emissions control responsibility should be assigned to upwind states, and how much responsibility should be left to downwind states."[63] Stakeholders have challenged the legality of EPA's interpretations over the years.[64]

The regulatory actions and litigation have led to EPA's establishment of the current framework to address the Good Neighbor provision for ozone and particulate matter.[65] The framework establishes a screening threshold—interstate pollution that exceeds 1% of the NAAQS—to identify states with sources that may contribute significantly to air quality problems in downwind states.[66] Upwind states that exceed this threshold for interstate pollution are evaluated further— considering cost and air quality factors—to determine whether emission reductions are needed.[67] EPA has clarified that it generally uses this framework to determine what constitutes a significant contribution when evaluating Good Neighbor SIPs and when evaluating a 126(b) petition.[68] See "Framework to Assess Good Neighbor Provision" for detailed discussion of the framework.

PROGRAMS ADDRESSING INTERSTATE POLLUTION

Pursuant to the CAA, EPA and states have implemented various market-based programs that target regional emissions of SO_2 and NO_x from power plants. One type of market-based program, known as emissions trading, sets a limit (or "cap") on total

[63] EPA, "Federal Implementation Plans," 76 *Federal Register* 48248.

[64] For a summary of how litigation has influenced implementation of the Good Neighbor provision, see CRS Report R43699, *Key Historical Court Decisions Shaping EPA's Program Under the Clean Air Act*, by Linda Tsang.

[65] EPA, "Federal Implementation Plans," 76 *Federal Register* 48208 (see in particular Section VI). For brief summary of the framework, see Peter Tsirigotis, Director, EPA Office of Air Quality Planning and Standards, memorandum to Regional Air Division Directors, EPA Regions 1-10, March 27, 2018, https://www.epa.gov/sites/production/files/2018- 03/documents/transport_memo_03_27_18_1.pdf.

[66] EPA, "Federal Implementation Plans," 76 *Federal Register* 48248. Under this approach, 1% of the 2015 ozone NAAQS (70 ppb) equals 0.7 ppb.

[67] EPA, "Cross-State Air Pollution Rule Update for the 2008 Ozone NAAQS," 81 *Federal Register* 74507-74508.

[68] EPA stated, "While either provision may be applied to address interstate transport, they are also closely linked in that a violation of the prohibition in CAA Section 110(a)(2)(D)(i) is a condition precedent for action under CAA Section 126(b) and, critically, that significant contribution to nonattainment and interference with maintenance are construed identically for purposes of both provisions (since the identical terms are naturally interpreted as meaning the same thing in the two linked provisions)." See EPA, "Response to Clean Air Act Section 126(b) Petitions from Delaware and Maryland," 83 *Federal Register* 26675.

emissions within a defined geographic area or economic sector and requires covered entities to surrender an allowance for each unit— typically a ton—of emissions.[69] Such programs are also known as "cap-and-trade."

Under an emissions cap, covered entities with relatively low emission-reduction costs have a financial incentive to reduce emissions because they can sell unused allowances to entities that face higher costs to reduce their facility emissions. The requirements vary by each program. For example, policymakers may decide to distribute the emission allowances to covered entities at no cost (based on, for example, previous years' emissions), sell the allowances (e.g., through an auction), or use some combination of these strategies. In addition, some programs may permit covered entities to "bank" or save surplus allowances for future use while others may not.

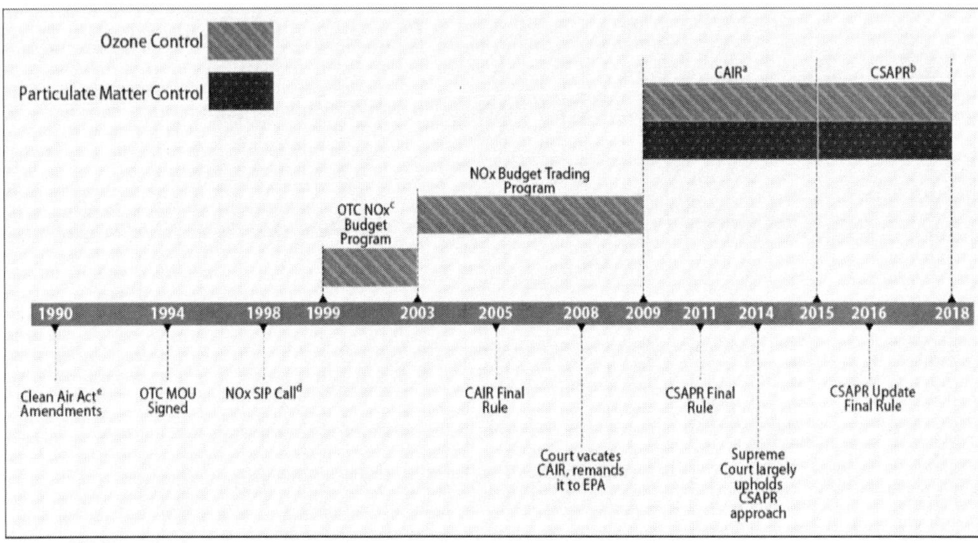

Source: CRS analysis.

Notes: The program dates in the figure represent the timeline for implementation and not enactment of statutory mandates. In addition, the Acid Rain Program, which seeks to reduce power sector emissions that cause acid rain (i.e., SO_2 and NO_x), was established under the 1990 CAA Amendments and remains in effect today.

[a] CAIR refers to the Clean Air Interstate Rule.

[b] CSAPR refers to the Cross State Air Pollution Rule.

[c] OTC NO_x refers to the Ozone Transport Commission Nitrogen Oxide Budget Program.

[d] NO_x SIP Call refers to the Nitrogen Oxide State Implementation Plan Call.

[e] Clean Air Act (CAA) of 1970 initially mandated the "Good Neighbor" provision, which was subsequently amended under the 1977 CAA and the 1990 amendments.

Figure 1. Implementation Timeline of Regional Ozone and Particulate Matter Control Programs.

[69] EPA, *What Is Emissions Trading?*, https://www.epa.gov/emissions-trading-resources/what-emissions-trading.

The remainder of this section presents a brief history of the interstate transport programs implemented prior to 2015, given their cumulative impact on regional emission reductions. The "Cross State Air Pollution Rule" section then provides more detail about CSAPR, the current emissions trading program intended to limit interstate transport of power sector SO_2 and NO_x emissions. *Figure 1* summarizes the timeline of the regional programs for ozone and particulate matter control

Acid Rain Program

EPA established the Acid Rain Program (ARP) under Title IV of the 1990 CAA Amendments to reduce power sector emissions that cause acid rain.[70] Specifically, the ARP targets SO_2 emissions through cap-and-trade and addressed NO_x emissions through an emissions-rate-based program.[71] Since its inception over two decades ago, the ARP has achieved notable reductions in these regional pollutants at lower-than-predicted costs. The market-based program also served as the basis for subsequent programs addressing interstate pollution.[72]

Under Title IV of the CAA, EPA implemented the ARP in two phases. The first phase—1995 to 1999—included 110 high-emitting coal-fired power plants, which had been identified in the statute and spanned 21 eastern and midwestern states.[73] The second phase began in 2000 and included more coal-fired power plants as well as those firing oil and natural gas, accounting for nearly all fossil-fueled power plants in the lower 48 states.[74] EPA set the annual SO_2 emissions cap at 9.97 million allowances in 2000 and decreased it in subsequent years. The ARP remains in effect today. The annual SO_2 cap—8.95 million tons of SO_2 per year—has not changed since 2010. The annual cap is roughly half of the SO_2 emitted by the power sector in 1980.[75]

---

[70] CAA §401; 42 U.S.C. §7651.

[71] CAA Title IV set goals for the year 2000 of reducing annual SO_2 emissions by 10 million tons from 1980 levels and reducing annual NO_x emissions by 2 million tons, also from 1980 levels. CAA §401(b); 42 U.S.C. §7651(b). See also CRS Report RL30853, *Clean Air Act: A Summary of the Act and Its Major Requirements*, by James E. McCarthy.

[72] According to EPA's *2016 Progress Report*, covered sources in the ARP emitted 1.5 million tons of SO_2 on an annual basis—a 91% reduction below 1990 levels—and 1.2 million tons of NO_x on an annual basis—an 81% reduction below 1990 levels. See EPA, *2016 Progress Report: Cross-State Air Pollution Rule and Acid Rain Program*, "Program Basics," 2016, pp. 7-8, https://www3.epa.gov/airmarkets/progress/reports/pdfs/2016_full_report.pdf. For additional discussion about the results of the ARP, see Gabriel Chan et al., *The SO_2 Allowance Trading System and the Clean Air Act Amendments of 1990: Reflections on Twenty Years of Policy Innovation*, Harvard Environmental Economics Program, January 2012, https://heep.hks.harvard.edu/files/heep/files/ SO2-brief_digital_final.pdf.

[73] CAA §404(e), Table A; 42 U.S.C. §7651(c)(e), Table A.

[74] EPA, *Acid Rain Program*, SO_2 Reductions, https://www.epa.gov/airmarkets/acid-rain-program. See also Richard Schmalensee and Robert Stavins, "Lessons Learned from Three Decades of Experience with Cap and Trade," *Review of Environmental Economics and Policy*, vol. 11, no. 1 (Winter 2017), p. 61.

[75] EPA, *2016 Progress Report*, p. 8. EPA reports the cap in units of allowances: The annual SO_2 cap is 8.95 million allowances. The CAA defines each allowance as one ton. See definition of *allowance* in CAA §402(3); 42 U.S.C. §7651(a)(3).

EPA distributed SO_2 allowances based on statutory formulas and accounted for historical emission rates and fuel consumption.[76] The "existing" power plant units—those in operation prior to November 15, 1990—received allowances for free.[77] The "new" power plant units—those commencing operations after November 15, 1990—generally did not receive free allowances and had to purchase them on the market.[78] At the end of each year, covered power plants have to surrender one allowance for each ton of SO_2 emitted. Unused allowances can either be sold or banked for use in later years.[79] The market value of unused allowances, therefore, serves as an incentive for power plants to "reduce emissions at the lowest cost."[80]

The NO_x portion of the ARP does not involve cap-and-trade but follows a more traditional regulatory approach. It is implemented through boiler-specific NO_x emission rates.[81] This program has provided power plants with some compliance flexibility, for example, by allowing the use of emissions rate averaging plans for units under common control, provided they meet certain conditions.[82] According to one analysis of the ARP, the NO_x portion "helped demonstrate the cost-effectiveness of NO_x controls," and by 2000, it "encouraged the installation of advanced NO_x combustion controls, such as low-NO_x burners, and the development of new power plant designs with lower NO_x emission rates."[83] While the ARP reduced SO_2 and NO_x emissions from the power sector, additional reductions were needed to meet ambient air quality standards under the CAA and to address the statute's Good Neighbor provision.[84] For example, in 1997, EPA revised the NAAQS for ozone and particulate matter—which are formed by SO_2 and NO_x—to be more stringent.[85] The next sections summarize some of the programs designed to achieve these reductions.

Ozone Control: Regional NO_x Programs

Ozone control strategies had focused on VOC emissions until the mid-1990s, when market-based programs began targeting another ozone precursor, NO_x, given its "important

[76] CAA §403; 42 U.S.C. §7651(b). EPA elaborated on and codified the distribution criteria at 40 C.F.R. 73.10.

[77] For statutory definition, see CAA §402(8); 42 U.S.C. §7651(a)(8). For statutory direction on distribution of SO_2 allowances, see CAA §403; 42 U.S.C. §7651(b).

[78] New power plants that met statutory criteria could receive some allowances from EPA. For statutory definition, see CAA §402(10); 42 U.S.C. §7651(a)(10). For statutory direction on distribution of SO_2 allowances to new units, see CAA §403(e); 42 U.S.C. §7651(b)(e).

[79] For more implementation details, see EPA, *Acid Rain Program*.

[80] Chan et al., *The SO_2 Allowance Trading System*, p. 4.

[81] EPA, *Acid Rain Program*, Overview.

[82] EPA, *2016 Progress Report*, p. 10.

[83] Sam Napolitano et al., "The U.S. Acid Rain Program: Key Insights from the Design, Operation, and Assessment of a Cap-and-Trade Program," *Electricity Journal*, vol. 20, no. 7 (August/September 2007), p. 50.

[84] Chan et al., *The SO_2 Allowance Trading System*, p. 17.

[85] For the 1997 ozone NAAQS revision, see EPA, "National Ambient Air Quality Standards for Ozone," 62 *Federal Register* 38856, July 18, 1997. For the 1997 particulate matter NAAQS revision, see EPA, "National Ambient Air Quality Standards for Particulate Matter," 62 *Federal Register* 38652, July 18, 1997.

role ... in ozone formation and transport."[86] Specifically, two regional trading programs were implemented between 1999 and 2009 to address ozone by reducing NO_x emissions. The first one, the "Ozone Transport Commission NO_x Budget Program," was in effect between 1999 and 2002. It was then replaced by the second program, the "NO_x Budget Trading Program," which ran until 2009.

Ozone Transport Commission NO_x Budget Program

The OTC—a multistate organization established under the 1990 CAA Amendments to advise EPA on ozone transport issues—developed the NO_x Budget Program and implemented it through a Memorandum of Understanding (MOU) with nearly all of the OTC states.[87] The OTC NO_x Budget Program set a regional budget (i.e., cap) on NO_x emissions from electric utilities and large industrial boilers during the "ozone season" (May through September), which is the time of year weather conditions are most favorable for ozone formation.[88] Under the MOU, states "were responsible for adopting regulations, identifying sources, allocating NO_x allowances, and ensuring compliance," while EPA was "responsible for approving the states' regulations and tracking allowances and emissions."[89]

Before emissions trading began under the OTC NO_x Budget Program, sources under the ARP were required to meet CAA emission rate standards that were in effect at that time. Sources could not emit above the NO_x level expected if using Reasonably Available Control Technology.[90] Next, the cap-and-trade program began in 1999 and ran until 2002, at which point the OTC NO_x Budget Program was effectively replaced by the NO_x Budget Trading Program (see next section).

In 2002, the sources participating in the OTC NO_x Budget Program reduced ozone season NO_x emissions 60% below 1990 baseline levels.[91] Despite the NO_x reductions in the Northeast, many northeastern and mid-Atlantic states were unable to meet a statutory

---

[86] EPA and OTC, *NOx Budget Program*, 1999-2002 Progress Report, March 2003, p. 1, https://www.epa.gov/sites/production/files/2015-08/documents/otcreport.pdf.

[87] The OTC is comprised of the governors of the 12 states in the OTR (Connecticut, Delaware, Maine, Maryland, Massachusetts, New Hampshire, New Jersey, New York, Pennsylvania, Rhode Island, Vermont, and Virginia) and the mayor of the District of Columbia. CAA §184; 42 U.S.C. §7511(c). All states in the OTC signed the MOU except for Virginia. EPA and OTC, *NO_x Budget Program*, p. 1. For the MOU, see https://otcair.org/upload/Documents/Memorandums/att2.htm.

[88] EPA and OTC, *NO_x Budget Program*, pp. 1-3

[89] EPA, *Ozone Transport Commission NO_x Budget Program*, https://www.epa.gov/airmarkets/ ozone-transport-commission- no_x -budget-program.

[90] The NO_x component of the ARP—boiler-specific NO_x emission rates on an annual basis—was in effect at the same time as the ozone NO_x trading programs, which addressed NO_x emissions in the summer (i.e., ozone season). The ARP NO_x program remains in effect today and applies to "ARP-affected units that are older, historically coal-fired boilers." EPA, *2016 Progress Report*, p. 15. See also EPA and OTC, *NO_x Budget Program*, p. 4. For summary and detailed analysis of the OTC NO_x Budget Program, see Dallas Burtraw and Sarah Jo Szambelan, *U.S. Emissions Trading Markets for SO_2 and NO_x*, Resources for the Future, October 2009, pp. 21-24, http://www.rff.org/files/sharepoint/WorkImages/ Download/RFF-DP-09-40.pdf.

[91] EPA and OTC, *NO_x Budget Program*, p. 6.

deadline to attain the one-hour ozone NAAQS. EPA concluded that these areas had not met this statutory deadline largely because of ozone transport from upwind areas.[92]

NO_x Budget Trading Program

The NO_x Budget Trading Program (NBP) effectively replaced the OTC NO_x Budget Program and was implemented between 2003 and 2009. The NBP encompassed a wider geographic area than the OTC NO_x Budget Program and targeted NO_x reductions from electric utilities and nonutility sources (e.g., large industrial boilers).[93] EPA established the NBP under the NO_x SIP Call, which required a number of eastern and midwestern states, plus the District of Columbia, to revise their SIPs to address regional ozone transport.[94] The NO_x SIP Call set a NO_x ozone season budget for each state and required upwind states to adopt SIPs that would reduce NO_x emissions to a level that would meet the budgets.

In the NO_x SIP Call, EPA observed that "virtually every nonattainment problem is caused by numerous sources over a wide geographic area," leading the agency to conclude that "the solution to the problem is the implementation over a wide area of controls on many sources, each of which may have a small or unmeasurable ambient impact by itself."[95] Ultimately, EPA expected that this would "eliminat[e] the emissions that significantly contribute to nonattainment or interference with maintenance of the ozone NAAQS in downwind states."[96]

EPA based the NO_x SIP Call in part on recommendations from the Ozone Transport Assessment Group (OTAG), a group created by EPA and the 37 easternmost states.[97] Of most relevance, OTAG recommended strategies to reduce NO_x emissions from utilities as well as large and medium nonutility sources in a trading program.[98]

EPA accounted for the cost of NO_x controls when establishing the NO_x budgets.[99] EPA identified cost-effective reductions in the electric utility and nonutility source sectors. These control strategies informed the establishment of the NO_x emission budgets.[100] EPA did not identify cost-effective controls in other sectors—namely, area sources (i.e., nonmobile sources

---

92 EPA, "Finding of Significant Contribution," 63 *Federal Register* 57361.

93 *Nonutility sources* refers to industrial sources that produce electricity or steam primarily for internal use. Examples include boilers and turbines at manufacturing facilities, such as paper mills and petroleum refineries. See EPA, *The NO_x Budget Program: 2008 Emission, Compliance, and Market Analyses*, July 2009, p. 3, https://www.epa.gov/sites/ production/files/2015-08/documents/nbp_2008_ecm_analyses.pdf.

94 EPA, "Finding of Significant Contribution," 63 *Federal Register* 57356.

95 EPA, "Finding of Significant Contribution," 63 *Federal Register* 57377.

96 EPA, "Response to Clean Air Act Section 126(b) Petitions from Delaware and Maryland," 83 *Federal Register* 26668.

97 The westernmost OTAG states ranged from North Dakota to Texas, with all remaining OTAG states located eastward of these states. For a map of OTAG states, see EPA, *Supplemental Ozone Transport Rulemaking Regulatory Analysis*, April 7, 1998, p. 1-2, https://archive.epa.gov/ttn/ozone/web/pdf/supregan.pdf.

98 EPA, "Finding of Significant Contribution," 63 *Federal Register* 57358-57359.

99 In 2000, the D.C. Circuit held in *Michigan v. EPA* that nothing in Section 110(a)(2)(D) bars EPA from considering costs. For additional information, see CRS Report R43699, *Key Historical Court Decisions Shaping EPA's Program Under the Clean Air Act*, by Linda Tsang.

100 EPA, "Finding of Significant Contribution," 63 *Federal Register* 57405.

that emit less than 100 tons of NO_x per year),[101] nonroad engines (i.e., mobile sources that do not operate on roads and highways, such as engines used to power snowmobiles, chainsaws, or lawnmowers),[102] or highway vehicles. Under the NO_x SIP Call, states could require their sources to comply with the emissions budget or participate in a regional cap-and-trade program. EPA developed a model rule for a regional emissions trading program—known as the NO_x Budget Trading Program—to assist states interested in the trading option. All of the jurisdictions—20 states and the District of Columbia—adopted the NBP into their SIPs and participated in the NBP.[103]

In 2008, NBP emissions were 9% below the 2008 cap, representing a 75% reduction compared to 1990 baseline levels.[104] This also represented a 62% reduction below a 2000 baseline, which accounted for emission reductions that occurred under the 1990 CAA Amendments before implementation of the NBP.[105]

EPA observed that ozone season NO_x emissions decreased each year between 2003 and 2008 and attributed these reductions in part to the installation of NO_x controls.[106] The agency noted that emissions vary year-to-year due to variables such as weather, electricity demand, and fuel costs. For example, EPA attributed the NO_x reductions between 2007 and 2008 primarily to lower electricity demand. However, analysis of the entire NBP period—2003 to 2008—shows a reduction in ozone season NO_x emissions despite a slight increase in demand for electricity. EPA reported that the average NO_x emission rate for the 10 highest electricity demand days (i.e., hot days when use of air conditioning is high) decreased in each year of the NBP. This metric for peak electricity days was 44% lower in 2008 compared to 2003.[107]

EPA reported that ozone concentrations decreased by 10% between the years 2002 and 2007 across all states participating in the NBP.[108] EPA also observed a "strong association between areas with the greatest NO_x emission reductions from NBP sources and downwind

---

[101] EPA, "Finding of Significant Contribution," 63 *Federal Register* 57498. Examples of area sources of NO_x emissions include wildfires and residential water heaters. EPA, "Finding of Significant Contribution and Rulemaking for Certain States in the Ozone Transport Assessment Group Region for Purposes of Reducing Regional Transport of Ozone," 62 *Federal Register* 60353, November 7, 1997.

[102] Additional examples of nonroad engines are those used for agricultural, construction, and mining equipment as well as aircraft and train engines. EPA, "Finding of Significant Contribution and Rulemaking for Certain States in the Ozone Transport Assessment Group Region for Purposes of Reducing Regional Transport of Ozone," 62 *Federal Register* 60359.

[103] The NO_x SIP Call included 22 states and the District of Columbia, but a later ruling removed one state (Wisconsin) and parts of two other states (Missouri and Georgia). The court left requirements for the remaining 19 states in place. See *Michigan v. EPA*, 213 F.3d 663 (D.C. Cir. 2000) *cert. denied* 532 U.S. 904 (2001). See also EPA, *2016 Program Progress*, "Program Basics."

[104] EPA, *The NOx Budget Program: 2008 Emission, Compliance, and Market Analyses*, July 2009, p. 4, https://www.epa.gov/sites/production/files/2015-08/documents/nbp_2008_ecm _analyses.pdf.

[105] EPA, *The NOx Budget Program*, p. 4.

[106] EPA, *The NOx Budget Program*, pp. 5-6.

[107] EPA, *The NOx Budget Program*, p. 6.

[108] Based on ozone concentrations averaged over an eight-hour time interval and after adjusting for meteorological conditions. See EPA, *NOx Budget Trading Program: Compliance and Environmental Results*, December 2008, p. 44, https://www.epa.gov/sites/production/ files/2015-08/documents/2007-nbp-report.pdf.

monitoring sites measuring the greatest improvements in ozone."[109] Progress notwithstanding, some NBP areas remained in nonattainment status with the ozone NAAQS as the NBP program concluded by the end of 2008.[110]

Ozone and PM Control: Regional SO_2 and NO_x Trading Programs

In 2005, EPA determined that interstate transport of SO_2 and NO_x contributed significantly to ozone and PM2.5 nonattainment.[111] Specifically, EPA found that (1) interstate transport of NO_x from 25 states and the District of Columbia contributed significantly to nonattainment, or interfered with maintenance, of the 1997 eight-hour ozone NAAQS; and (2) interstate transport of SO_2 and NO_x from 23 states and the District of Columbia contributed significantly to nonattainment, or interfered with maintenance, of the 1997 PM2.5 NAAQS.[112] To address these findings, EPA promulgated a rule that applied to 28 eastern states and the District of Columbia.[113] This rulemaking is known as the Clean Air Interstate Rule (CAIR).

A legal challenge, however, vacated and remanded CAIR to EPA.[114] CAIR remained in effect while EPA responded to the court decision and developed a new regional program addressing air transport, known as CSAPR. CSAPR replaced CAIR on January 1, 2015, and remains in effect today.[115] The remainder of this section discusses each program in turn.

---

[109] Based on ozone concentrations averaged over an eight-hour time interval and after adjusting for meteorological conditions. See EPA, *NOx Budget Trading Program: Compliance and Environmental Results*, p. 44.

[110] EPA, *The NOx Budget Program: 2008 Emission, Compliance, and Market Analyses*, July 2009, p. 6, https://www.epa.gov/sites/production/files/2015-08/documents/nbp_2008_ ecm_analyses.pdf. The NBP was effectively replaced by the ozone season NOx component of the Clean Air Interstate Rule. See EPA, *2016 Program Progress*, p. 8.

[111] EPA identified the geographic areas and interstate air emissions transport through a series of modeling runs. EPA also used several criteria to determine what constituted a "significant contribution" for ozone nonattainment and for PM2.5 nonattainment. EPA changed this approach in CSAPR, which replaced CAIR. The CSAPR approach is discussed later in this report.

[112] EPA, "Rule to Reduce Interstate Transport of Fine Particulate Matter and Ozone (Clean Air Interstate Rule); Revisions to Acid Rain Program; Revisions to the NOx SIP Call; Final Rule," 70 *Federal Register* 25167, May 12, 2005.

[113] EPA, "Rule to Reduce Interstate Transport of Fine Particulate Matter and Ozone (Clean Air Interstate Rule)," 70 *Federal Register* 25170. The totals reported in this paragraph include Minnesota, which EPA included in the CAIR final rule. EPA subsequently promulgated a final rule to stay the effectiveness of CAIR and associated CAIR FIP requirements for Minnesota. More recent publications may therefore exclude Minnesota from the total number of states that were subject to CAIR. For Final Stay of CAIR and CAIR FIP for Minnesota, see EPA, "Administrative Stay of Clean Air Interstate Rule for Minnesota; Administrative Stay of Federal Implementation Plan to Reduce Interstate Transport of Fine Particulate Matter and Ozone for Minnesota," 74 *Federal Register* 56721, November 3, 2009.

[114] *North Carolina v. EPA*, 550 F.3d 1176 (D.C. Cir. 2008). See also CRS Report R43699, *Key Historical Court Decisions Shaping EPA's Program Under the Clean Air Act*, by Linda Tsang.

[115] EPA, *2016 Program Progress*, p. 9.

Clean Air Interstate Rule

CAIR established a regional cap-and-trade program to reduce power sector SO_2 and NO_x emissions. Specifically, CAIR established emission budgets for each of the 28 states as well as a model rule for a multi-state cap-and-trade program in the power sector.[116] Under CAIR, states could achieve their emission budgets by requiring their sources to participate in the cap-and-trade program.

CAIR set three emissions caps: Two were annual emissions caps to limit SO_2 and NO_x as precursor emissions to PM2.5, and the third was an ozone season cap limiting NO_x as a precursor emission to ozone. The annual NO_x and seasonal NO_x caps were implemented as the "CAIR NO_x annual" and "CAIR ozone season NO_x" programs, respectively, in 2009. The SO_2 emissions cap was implemented as the "CAIR SO_2 annual" program in 2010.[117]

The scope of CAIR differed from prior NO_x trading programs. Whereas the NBP had included both electric generators and nonutility industrial sources (e.g., boilers and turbines), CAIR focused only on electric generators. As previously noted, OTAG's recommendations for the NO_x SIP Call included NO_x controls for medium and large nonutility stationary sources as well as electric generating units. While nonutility sources emit both NO_x and SO_2, EPA did not require NO_x and SO_2 reductions from these sources under CAIR. EPA concluded that it needed more reliable emissions data and better information about control costs to require reductions from nonutility sources in CAIR. Specifically, EPA stated that it lacked information about the costs to integrate NO_x and SO_2 controls at nonutility sources and therefore could not determine whether such controls would qualify as "highly cost-effective" under CAIR.[118]

Some stakeholders disagreed with this conclusion, noting that EPA had cost information from the NO_x SIP Call. EPA responded that the geographic scope of the NO_x SIP Call differed somewhat from CAIR, and therefore it had limited emissions data about nonutility sources in CAIR states that were outside of the NO_x SIP Call. In addition, EPA expected that projected NO_x and SO_2 emissions from nonutility sources were "significantly lower than projected" emissions from electric generators.[119] EPA concluded that states would be better positioned to "make decisions regarding any additional control requirements for [non-utility] sources."[120]

CAIR was challenged in court. The U.S. Court of Appeals for the District of Columbia concluded that CAIR was flawed, finding among other things that the CAIR trading

---

[116] EPA, "Rule to Reduce Interstate Transport of Fine Particulate Matter and Ozone (Clean Air Interstate Rule)," 70 *Federal Register* 25162.

[117] EPA, *2016 Program Progress*, pp. 10-11.

[118] EPA, "Rule to Reduce Interstate Transport of Fine Particulate Matter and Ozone (Clean Air Interstate Rule)," 70 *Federal Register* 25214-25215.

[119] EPA projected that in 2010, under base case conditions, electric generator emissions would contribute 70% of SO_2 (and 25% NOx) in the CAIR region compared to 15% of SO_2 (and 16% NOx) from nonelectric generator boilers and turbines. EPA, "Rule to Reduce Interstate Transport of Fine Particulate Matter and Ozone (Clean Air Interstate Rule)," 70 *Federal Register* 25214.

[120] EPA, "Rule to Reduce Interstate Transport of Fine Particulate Matter and Ozone (Clean Air Interstate Rule)," 70 *Federal Register* 25215.

program did not assure some "measurable" emission reduction in each upwind state.[121] The court reasoned that the "[e]missions reduction by the upwind states *collectively* was not enough to satisfy Section 110(a)(2)(D)."[122] The court ultimately remanded CAIR to EPA in December 2008, allowing CAIR to remain in effect while EPA developed a replacement rule.[123] The CAIR programs for NO_x (annual and ozone season) began in 2009 and the CAIR SO_2 program began in 2010. The programs continued through the end of 2014.[124]

Cross State Air Pollution Rule

In 2011, EPA promulgated CSAPR to address the court's concerns regarding CAIR.[125] CSAPR implementation began in 2015—replacing CAIR—and it remains in effect today. Similar to CAIR, CSAPR aims to reduce ozone and $PM_{2.5}$ interstate transport. As shown in Figure 2, CSAPR requires 27 states to reduce SO_2 emissions, annual NO_x emissions, and/or ozone season NO_x emissions from the power sector.[126] Specifically, CSAPR sets annual SO_2, annual NO_x, and ozone-season NO_x budgets for the covered states and allows states to determine how they will achieve those budgets, including the option of emissions trading.

CSAPR differs from CAIR in other ways, though, and introduced a new approach to measuring a significant contribution under Section 110(a)(2)(D). EPA had previously relied on a regional analysis of significant contributions (e.g., in CAIR and the NO_x SIP Call).[127] As previously noted, the D.C. Circuit found the regional approach flawed in a ruling on CAIR.[128] As a result, EPA used state-specific information under CSAPR to determine significant contributions at the state level. After various legal challenges, the approach used in CSAPR remains in effect today. EPA has determined that it can use this framework to assess the Good Neighbor provision each time it revises the relevant NAAQS.[129]

---

[121] *North Carolina v. EPA*, 531 F.3d 896, 908 (D.C. Cir. 2008). CRS Report R43699, *Key Historical Court Decisions Shaping EPA's Program Under the Clean Air Act*, by Linda Tsang.

[122] CRS Report R43699, *Key Historical Court Decisions Shaping EPA's Program Under the Clean Air Act*, by Linda Tsang.

[123] Though initially vacating the CAIR rule, the D.C. Circuit revised its opinion some months later, leaving CAIR in effect "until it is replaced by a rule consistent with our opinion." *North Carolina v. EPA*, 550 F.3d 1176, 1178 (D.C. Cir. 2008). See also CRS Report R43699, *Key Historical Court Decisions Shaping EPA's Program Under the Clean Air Act*, by Linda Tsang.

[124] EPA, 2016 Program Progress, p. 11.

[125] EPA, "Federal Implementation Plans," 76 Federal Register 48207.

[126] CSAPR covers 27 states as of July 2018. Some changes have been made since EPA promulgated the 2011 CSAPR rulemaking. For list of CSAPR states, see EPA, "States That Are Affected by the Cross-State Air Pollution Rule (CSAPR)," https://www.epa.gov/csapr/ states-are-affected-cross-state-air-pollution-rule-csapr.

[127] EPA, "Federal Implementation Plans," 76 *Federal Register* 48248.

[128] *North Carolina v. EPA*, 550 F.3d 1176 (D.C. Cir. 2008). See also CRS Report R43699, *Key Historical Court Decisions Shaping EPA's Program Under the Clean Air Act*, by Linda Tsang.

[129] EPA, "Cross-State Air Pollution Rule Update," 81 *Federal Register* 74507.

Source: EPA (2018), "Map of States Covered by CSAPR," https://www.epa.gov/ airmarkets/map-states-coveredcsapr.

Note: Alaska and Hawaii (not shown) are not covered under the Acid Rain Program (ARP), which applies to power plants in the contiguous United States, nor are they covered under the Cross State Air Pollution Rule (CSAPR).

Figure 2. States Covered by CSAPR.

Framework to Assess Good Neighbor Provision

EPA developed a multi-step framework to assess states' Good Neighbor obligations and determine each state's significant contribution in CSAPR. First, EPA conducted air quality modeling to project "downwind air quality problems"—that is, it identified downwind monitoring receptors expected to have difficulty attaining or maintaining the NAAQS.[130] Next, EPA identified the links between upwind states and the downwind air quality monitoring sites with projected attainment or maintenance difficulties. EPA then identified which of these linked upwind states "contribute at least one percent of the relevant NAAQS" at the downwind sites.[131] The agency next assessed the cost-effectiveness of emission control measures and air quality factors to determine whether states exceeding this threshold made significant contributions or interfered with maintenance of a NAAQS in a downwind state. That is, EPA determined that an upwind state contributes significantly to a nonattainment or interference with maintenance of a

[130] For a summary of the process, see EPA, "Federal Implementation Plans," 76 *Federal Register* 48211. See also Tsirigotis, pp. 2-3.

[131] EPA, "Federal Implementation Plans," 76 *Federal Register* 48248.

NAAQS if it produced more than 1% of NAAQS concentration in at least one downwind state and if this pollution could be mitigated using cost-effective measures.

EPA modified the way it considered costs under CSAPR. Whereas EPA had previously based "significant contribution" on the emissions that "could be removed using 'highly cost effective' controls," the agency accounted for both cost and air quality improvement to measure significant contributions under CSAPR.[132] In CSAPR, EPA (1) quantified each state's emission reductions available at increasing costs per ton ("cost thresholds"), (2) evaluated the impact of upwind reductions on downwind air quality, and (3) identified the cost thresholds providing "effective emission reductions and downwind air quality improvement."[133]

The last step of the Good Neighbor assessment framework requires the adoption of "permanent and enforceable measures needed to achieve" the emission reductions.[134] EPA implemented this step through its promulgation of FIPs, giving states the option to replace the FIP with a SIP.[135] The FIPs specified the emission budgets for each state, reflecting the required SO_2 and NO_x reductions from power plants in the state, and established the trading programs as each state's remedy to meet the emissions budgets.[136]

Legal challenges, which eventually reached the Supreme Court, delayed CSAPR implementation.[137] The Court largely upheld EPA's approach, holding that EPA's consideration of cost in establishing states' emission budgets was a "permissible construction of the statute."[138]

CSAPR Emissions Trading Programs

In response to the CAIR litigation, EPA designed "air quality-assured interstate emission trading programs" to implement CSAPR.[139] The CSAPR trading programs allow for interstate trading but include provisions meant to ensure that all of the necessary reductions would occur in each individual state. Specifically, EPA stated that the CSAPR assurance provisions "ensure that no state's emissions ... exceed that specific state's budget plus the variability limit (i.e., the state's assurance level)."[140]

---

[132] EPA, "Federal Implementation Plans," 76 *Federal Register* 48248.

[133] EPA, "Federal Implementation Plans," 76 *Federal Register* 48248.

[134] Tsirigotis, p. 3.

[135] EPA, "Federal Implementation Plans," 76 *Federal Register* 48209.

[136] See EPA, "Federal Implementation Plans," 76 Federal Register 48210, and EPA, Regulatory Impact Analysis of the Cross-State Air Pollution Rule (CSAPR) Update for the 2008 National Ambient Air Quality Standards for Ground-Level Ozone, September 2016, p. ES-8, https://www3.epa.gov/ttn/ecas/docs/ria/transport_ria_final-csapr-update_2016- 09.pdf.

[137] *EPA v. EME* Homer City Generation, L.P., 134 S. Ct. 1584 (2014). For summary of the CSAPR legal challenges and the Supreme Court's ruling, see CRS Report R43699, *Key Historical Court Decisions Shaping EPA's Program Under the Clean Air Act*, by Linda Tsang.

[138] *EPA v. EME* Homer City Generation, L.P., 134 S. Ct. 1584, 1606–07 (2014).

[139] EPA, "Federal Implementation Plans," 76 *Federal Register* 48210.

[140] Each state's emissions budget accounts for the "inherent variability in power system operations." See EPA, "Federal Implementation Plans," 76 *Federal Register* 48212, 48265, 48303.

EPA established four interstate trading programs for affected power plants under CSAPR: two for annual SO_2, one for annual NO_x, and one for ozone-season NO_x.[141] These trading programs aim to help downwind areas attain the 1997 and 2006 annual PM2.5 NAAQS and the 1997 and 2008 ozone NAAQS. The first phase of CSAPR, which began in 2015, sought to address the 1997 and 2006 PM2.5 NAAQS as well as the 1997 ozone NAAQS. The second phase of CSAPR, referred to as the CSAPR Update, began in 2017 and has sought to address the 2008 ozone NAAQS.[142]

The total emissions budget for each CSAPR trading program equals the sum of the individual state budgets covered by that program. Affected power plants receive an allocation of allowances based on the emission budget for that trading program in the state. Each affected power plant must have an allowance to emit each ton of the relevant pollutant. It may comply with its allowance allocation by using control technologies to reduce emissions—and sell or bank any surplus allowances—or buy more allowances on the market.[143]

EPA's "CSAPR Update" rulemaking updated the ozone season NO_x program with respect to the 2008 ozone NAAQS.[144] Specifically, the CSAPR Update promulgated new FIPs for 22 states; 21 of these states were covered in the original CSAPR ozone season NO_x trading program.[145] The updated ozone season NO_x trading began in 2017 and largely replaced the original CSAPR ozone season NO_x trading program.[146] EPA concluded based on its modeling analysis that emissions from 10 of the states covered in the original CSAPR ozone season NO_x trading program "no longer significantly contribute to downwind nonattainment or interference with maintenance" of either the 1997 ozone NAAQS or the 2008 ozone NAAQS.[147] Various states and stakeholders have filed a petition for review of the CSPAR Update to the D.C. Circuit.[148]

---

[141] One SO_2 trading program—SO_2 Group 1—covers sources in states that need to make larger reductions to eliminate their significant contribution, while the second—SO_2 Group 2—includes sources in states that need to make smaller reductions. EPA, *2016 Program Progress*, p. 9.

[142] EPA, *2016 Program Progress*, p. 9.

[143] EPA clarified that while allowance trading—buying, selling, and banking—is allowed without restriction, there is a limit on the surrender of SO_2 allowances. A power plant in a Group 1 state can use only SO_2 allowances allocated to Group 1 states. Likewise, a power plant in a Group 2 state can use only Group 2 SO_2 allowances to cover SO_2 emissions. EPA, "Federal Implementation Plans," 76 *Federal Register* 48263-42864, 48284. See also EPA, *2016 Program Progress*, p. 9.

[144] EPA, "Cross-State Air Pollution Rule Update for the 2008 Ozone NAAQS," 81 *Federal Register* 74507.

[145] Kansas was not covered in the original CSAPR ozone season NOx program but was included in the CSAPR Update. See EPA, "Cross-State Air Pollution Rule Update for the 2008 Ozone NAAQS," 81 *Federal Register* 74506.

[146] The original ozone season CSAPR program was renamed "CSAPR NOx Ozone Season Group 1 Trading Program" and now includes just one state: Georgia. Georgia is not included in the CSAPR Update trading program, "CSAPR NOx Ozone Season Group 2." See EPA, "Determination Regarding Good Neighbor Obligations for the 2008 Ozone National Ambient Air Quality Standard," 83 *Federal Register* 31922, July 10, 2018; EPA, *Overview of the Cross-State Air Pollution Rule (CSAPR)*, https://www.epa.gov/csapr/overview-cross-state-air-pollution-rule-csapr; and EPA, "Cross-State Air Pollution Rule Update for the 2008 Ozone NAAQS," 81 *Federal Register* 74504.

[147] EPA, "Cross-State Air Pollution Rule Update for the 2008 Ozone NAAQS," 81 *Federal Register* 74507.

[148] *Wisconsin v. EPA*, No. 16-1406 (D.C. Cir. November 23, 2016).

CSAPR does not address the Good Neighbor provision with respect to either the 2012 revision to the PM2.5 NAAQS[149] or the 2015 revision[150] to the ozone NAAQS.[151] As of July 2018, states and EPA are in the process of evaluating interstate ozone transport with respect to the 2015 ozone NAAQS (see discussion under "Good Neighbor Determinations and the 2015 Ozone Standard").

Regarding the 2012 PM2.5 standard, a 2016 EPA analysis determined that "few areas in the United States" would "have problems attaining and maintaining the 2012 PM2.5 NAAQS due to the relatively small number and limited geographic scope of projected nonattainment and maintenance receptors."[152] EPA concluded that "most states will be able to develop good neighbor SIPs that demonstrate that they do not contribute significantly to nonattainment or interfere with maintenance of the 2012 PM2.5 NAAQS in any downwind state."[153] Currently, nine areas are designated nonattainment with the 2012 PM2.5 standard, four of which are located in two CSAPR states (Ohio and Pennsylvania). No areas are currently designated as maintenance with that standard.[154]

Results of Regional SO_2 and NO_x Trading Programs

Power sector SO_2 and NO_x emissions have declined since 2005. EPA has attributed most of these reductions to CAIR, which was in effect through the end of 2014.[155] The agency noted that other programs, such as state NO_x emission control programs, also contributed to the reductions in annual and ozone season NO_x achieved by 2016.[156] *Figure 3* illustrates the trend of declining emissions, showing that annual SO_2, annual NO_x, and ozone season NO_x decreased between 2009 (the first year of CAIR) and 2016 (the latest year for which the EPA Air Markets Program Data website reports emissions for all three programs).

EPA attributed the SO_2 reductions under CAIR/CSAPR and the ARP largely to the greater use of pollution control technologies on coal-fired power plant units and "increased generation at natural gas-fired units that emit very little SO_2 emissions."[157] As noted by the

---

[149] Referred to as the "2012" PM2.5 NAAQS revision because EPA signed the rule in late 2012. The rule was published in the *Federal Register* in early 2013. EPA, "National Ambient Air Quality Standards for Particulate Matter," 78 *Federal Register* 3086, January 15, 2013.

[150] EPA, "National Ambient Air Quality Standards for Ozone," 80 *Federal Register* 65292, October 26, 2015.

[151] EPA, "Cross-State Air Pollution Rule Update for the 2008 Ozone NAAQS," 81 *Federal Register* 74507. EPA stated it addressed the Good Neighbor provision for the 2012 PM2.5 NAAQS in a memorandum. See EPA, "Information on the Interstate Transport 'Good Neighbor' Provision for the 2012 Fine Particulate Matter National Ambient Air Quality Standards under Clean Air Act Section 110(a)(2)(D)(i)(I)," March 17, 2016, https://www.epa.gov/pm-pollution/information-interstate-transport-good-neighbor-provision-2012-fine-particulate-matter.

[152] Stephen D. Page, Director, EPA Office of Air Quality Planning and Standards, letter to EPA Regional Air Division Directors, Regions 1-10, March 17, 2016, p. 3, https://www.epa.gov/sites/production/files/2016-08/documents/goodneighbor-memo_implementation.pdf.

[153] Page, March 17, 2016, p. 7.

[154] As of June 30, 2018. The states with nonattainment areas for the 2012 PM2.5 standard are California, Idaho, Ohio, and Pennsylvania. See EPA, *Green Book: PM-2.5 (2012) Designated Area/State Information*, June 30, 2018, https://www3.epa.gov/airquality/ greenbook/kbtc.html.

[155] EPA, 2016 Program Progress, p. 18.

[156] EPA, 2016 Program Progress, pp. 24, 30.

[157] EPA, 2016 Program Progress, p. 19.

U.S. Energy Information Administration (EIA), nearly all SO_2 emissions from the electricity sector are associated with coal-fired generation.[158] EPA reported that the average SO_2 emissions rate for units subject to either the CSAPR or ARP decreased 81% compared to 2005 rates. Most of the reductions were from coal-fired units.[159]

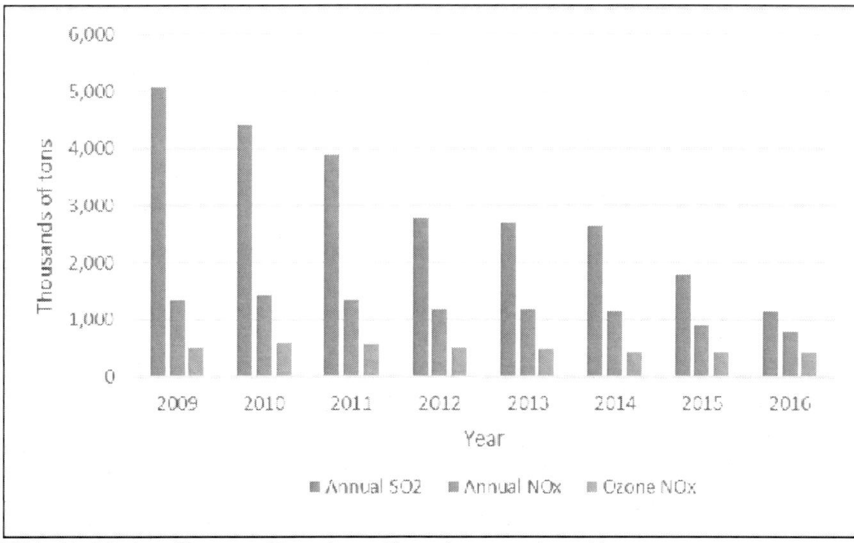

Source: EPA Air Markets Program Data, https://ampd.epa.gov/ampd/.

Note: CAIR was in effect from 2009 through the end of 2014 and was replaced by CSAPR on January 1, 2015.

Figure 3. CSAPR Sources Emission Trends, 2009-2016.

Analysis from EIA reveals a similar trend at the national level, suggesting that a combination of market and regulatory factors have contributed to SO_2 reductions. EIA reported a 73% reduction in national power sector SO_2 emissions from 2006 to 2015, which it described as "much larger" than the 32% reduction in coal-fired generation in that same period.[160] EIA attributed the national SO_2 reductions to (1) changes in the electricity generation mix (e.g., less coal-fired generation and more natural-gas-fired generation), (2) the installation of pollution control technologies at coal- and oil-fired plants (in particular, to comply with the Mercury and Air Toxics rule), and (3) lower use of the most-polluting power plants (e.g., retirements of coal-fired units).[161] Another EIA analysis reported that the eastern region of the United States—which includes all of the CSAPR states except

[158] EIA, "Sulfur Dioxide Emissions from U.S. Power Plants Have Fallen Faster Than Goal Generation," February 3, 2017, https://www.eia.gov/todayinenergy/detail.php?id=29812.

[159] EPA, 2016 Program Progress, p. 19.

[160] EIA, "Sulfur Dioxide Emissions."

[161] EIA, "Sulfur Dioxide Emissions." EIA stated that various factors influence the decision to retire a power plant and analyzed retirements from the past decade in a separate report. See EIA, "Almost All Power Plants That Retired in the Past Decade Were Powered by Fossil Fuels," January 9, 2018, https://www.eia.gov/todayinenergy/detail.php?id= 34452.

Texas—had the largest share of capacity retirements between 2008 and 2017 compared to the rest of the continental United States.[162]

In addition, emissions in 2016 were below the total emission budgets for each CSAPR trading program (see *Figure 4*).[163] EPA observed that this resulted in CSAPR allowance prices at the end of 2016 that "were well below the marginal cost for reductions projected at the time of the final rule [and that such prices] are subject, in part, to downward pressure from the available banks of allowances."[164]

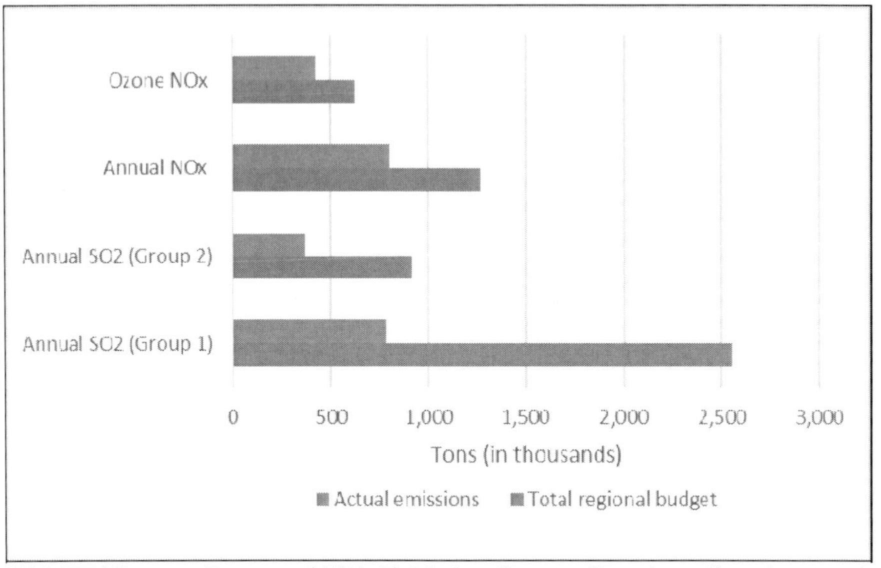

Source: EPA 2016 Progress Report and EPA Air Markets Program Data, https://ampd.epa.gov/ampd/.

Figure 4. Comparison of CSAPR Emissions Budgets and Actual Emissions in 2016.

EPA reported that preliminary data from the 2017 ozone season—the first CSAPR Update compliance period—show that ozone season NO_x emissions were below the total emission budget.[165]

Emission allowance prices are generally affected by a number of factors, including supply and demand, program design elements that influence supply and demand, and legal and regulatory uncertainty.[166] Analyses of ozone season NO_x highlight summer weather as a key factor (e.g., higher than average temperatures could lead to greater demand for electricity). Power sector compliance strategies (e.g., use of installed control technologies,

---

[162] EIA analyzed the lower 48 states of the United States and divided them into three regions: Western, Texas, and Eastern. See EIA, "Almost All Power Plants."

[163] EPA, 2016 Program Progress, pp. 24, 30.

[164] EPA, 2016 Program Progress, pp. 24, 30.

[165] In 2017, CSAPR Update sources emitted a total of 294,478 tons of ozone season NO_x, below the total budget of 316,464 tons. EPA, "Determination Regarding Good Neighbor Obligations," 83 Federal Register 31927.

[166] Allowance banking—saving allowances for use in future years—is an example of a program design feature that may influence allowance supply. For discussion about how legal and regulatory uncertainty has affected allowance prices in regional SO_2 and NO_x markets, see Burtraw and Szambelan, U.S. Emissions Trading Markets for SO_2 and NO_x.

switching to lower emitting fuels, or retiring higher emitting units) are also relevant to ozone season allowance prices.[167]

Recent allowance prices in the CSAPR Update trading program appear to be lower than the marginal cost to reduce ozone season NO_x emission. One brokerage firm reported that by May 2018—the start of the 2018 ozone season—NO_x allowance prices ranged from $150 to $175 per ton, suggesting that the availability of allowance prices at such low prices "could lead to some decisions not to run some pollution controls at maximum output. This would, in turn, lead to higher emissions."[168]

The brokerage firm reported the marginal cost of ozone season NO_x reductions to be about $300 per ton, though EPA considered higher marginal costs to develop the CSPAR Update emission budgets.[169] Specifically, EPA considered several cost thresholds—ranging from $800 per ton to $6,400 per ton—and based the CSAPR Update emission budgets on reductions that could be achieved at $1,400 per ton. EPA concluded that a $1,400 per ton threshold would maximize the incremental benefits—the emission reductions and corresponding downwind air quality improvements—compared to other marginal cost thresholds.[170] EPA identified NO_x control strategies at this cost threshold to include optimizing use of existing operational Selective Catalytic Reduction (SCR) controls, turning on existing but idled controls—for example, SCR that had not been used for several seasons—and installing advanced combustion controls, such as low-NO_x burners.[171]

EPA has reported improvements in air quality, attributing progress in part to the regional SO_2 and NO_x transport programs.[172] For example, 34 of the 36 areas in the eastern United States that were designated as nonattainment for the 1997 PM2.5 NAAQS now show concentrations below that standard based on 2014-2016 data.[173] In terms of ozone, all 92 of the eastern areas originally identified as nonattainment under the 1997 ozone standard now show concentrations below that standard based on 2014-2016 data.[174] The 2014-2016 monitoring data also showed that 17 of the 22 areas in the eastern United States that were originally designated as nonattainment with the 2008 ozone standard now have concentrations below that standard.[175]

---

[167] For example, see Evolution Markets, *Market Update: Cross State Air Pollution Rule*, March 21, 2018, p. 2, http://www.evomarkets.com/content/news/reports_28_report_file.pdf; and Energy Ventures Analysis, "First Ozone Season Under the CSAPR Update Rule Begins Amid Uncertain Market Conditions," May 31, 2017, https://www.evainc.com/energy-blog/first-csapr-ozone-season-underway/.

[168] Evolution Markets, *Market Update*, p. 2, http://www.evomarkets.com/content/news/reports_ 28_report_file.pdf.

[169] Evolution Markets, *Market Update*, p. 2.

[170] EPA, "Cross-State Air Pollution Rule Update for the 2008 Ozone NAAQS," 81 *Federal Register* 74508, 74543.

[171] EPA, "Cross-State Air Pollution Rule Update for the 2008 Ozone NAAQS," 81 *Federal Register* 74541, 74543. See also EPA, "Determination Regarding Good Neighbor Obligations," 83 *Federal Register* 31927.

[172] EPA, 2016 Program Progress, pp. 55-70.

[173] EPA had designated a total of 39 areas as nonattainment with respect to the 1997 $PM_{2.5}$ standard; 36 of these areas were located in the eastern United States. EPA, 2016 Program Progress, p. 67.

[174] EPA, 2016 Program Progress, p. 59.

[175] EPA, 2016 Program Progress, p. 59.

STATUS OF GOOD NEIGHBOR DETERMINATIONS FOR OZONE STANDARDS

As previously noted, revisions to the NAAQS trigger the SIPs review process, through which EPA determines whether states have met their Good Neighbor obligations. EPA has not yet finalized its Good Neighbor determinations for either the 2008 revision or the 2015 revision to the ozone standards. The remainder of this section summarizes the status of EPA's Good Neighbor determinations under each standard.

Good Neighbor Determinations and the 2008 Ozone Standard

EPA first sought to address ozone transport with respect to the 2008 ozone standard in the 2016 CSAPR Update. Specifically, the CSAPR Update covered 22 states and promulgated FIPs with ozone season NO_x budgets for power plants.[176] EPA concluded at the time, however, that it could not determine whether the CSAPR Update fully addressed the Good Neighbor provision with respect to the 2008 ozone standard for 21 of the 22 covered states.[177] In other words, the 2016 CSAPR Update "did not fully satisfy the EPA's obligation to address the good neighbor provision requirements" for those 21 states.[178] EPA based its 2016 conclusion in part on the agency's projection of air quality problems at downwind monitors in 2017, even with implementation of the CSAPR Update. EPA found that 21 of the 22 CSAPR Update states would contribute "equal to or greater than 1 percent of the 2008 ozone NAAQS" to at least one nonattainment or maintenance monitor in 2017.[179]

Since then, EPA has updated its air quality modeling and, on June 29, 2018, proposed to determine that the CSAPR Update fully addresses 20 of the 21 remaining Good Neighbor obligations for the 2008 ozone standards.[180] As such, the agency has "proposed to determine that it has no outstanding, unfulfilled obligation under Clean Air Act Section 110(c)(1) to establish additional requirements for sources in these states to further reduce transported

---

[176] EPA, "Cross-State Air Pollution Rule Update for the 2008 Ozone NAAQS," 81 *Federal Register* 74504, 74540.

[177] Tennessee was the only state for which EPA found the CSAPR Update FIP fully addressed the Good Neighbor provision with respect to the 2008 ozone standard. See EPA, "Cross-State Air Pollution Rule Update for the 2008 Ozone NAAQS," 81 *Federal Register* 74504, 74540; and EPA, "Determination Regarding Good Neighbor Obligations," 83 *Federal Register* 31920.

[178] EPA, "Determination Regarding Good Neighbor Obligations," 83 *Federal Register* 31920.

[179] EPA's analysis projected nonattainment monitors in Connecticut, Texas, and Wisconsin in 2017 and maintenance-only monitors in Connecticut, Maryland, Michigan, New York, and Texas in 2017. EPA, "Determination Regarding Good Neighbor Obligations," 83 *Federal Register* 31922.

[180] Kentucky is the only one out of the group of 21 not covered by this proposed determination. EPA will address Kentucky in a separate rulemaking. EPA, "Determination Regarding Good Neighbor Obligations," 83 *Federal Register* 31915, 31923.

ozone pollution under" the CAA's Good Neighbor provision with respect to the 2008 ozone NAAQS.[181]

EPA based its proposed determination on the updated air quality modeling, which projected air quality in 2023—a longer analytical time frame than it used in the CSAPR Update.[182] The updated projections showed that in 2023, there would not be any nonattainment or maintenance monitors with respect to the 2008 ozone standard in the eastern United States.[183]

EPA's selection of a future analytic year is an important factor in the Good Neighbor determination.[184] The agency based its selection of 2023 on two primary factors: (1) the downwind attainment deadlines[185] and (2) the time frame required to implement emission reductions as "expeditiously as possible."[186] As of August 2018, the next attainment dates for the 2008 ozone standard are July 20, 2021 (for areas classified as "Serious" nonattainment) and July 20, 2027 (for areas classified as "Severe" nonattainment).[187]

The potential to "over-control" emissions was another factor that EPA identified as relevant to the selection of the analytic year. EPA described it as relevant given the agency's expectation that future emissions will decline through implementation of existing local, state, and federal programs and in light of holdings from the U.S. Supreme Court.[188] EPA stated that it considered both downwind states' obligation to attain the ozone standards "as expeditiously as possible" and EPA's "obligation to avoid unnecessary over-control of upwind state emissions."[189] EPA did not specify whether it expected separate agency actions that may affect ozone precursor emissions— such as changes in the mobile source program—to affect its projections for 2023.

---

181 EPA, "Determination Regarding Good Neighbor Obligations," 83 *Federal Register* 31916.

182 In the CSAPR Update, EPA selected the year 2017 because it was the last full ozone season from which data could be used to determine attainment with the 2008 ozone standard by July 20, 2018, the attainment date for nonattainment areas classified as moderate. In addition, EPA did not analyze ozone season NO_x reductions available after 2017 due to time constraints. Stephen D. Page, Director, EPA Office of Air Quality Planning and Standards, memorandum to EPA Regional Air Division Directors, Regions 1-10, October 27, 2017, p. 3, https://www.epa.gov/airmarkets/october-2017-memo-and-supplemental-information-interstate-transport-sips-2008-ozone-naaqs.

183 Page, October 27, 2017, p. 1, and EPA, "Determination Regarding Good Neighbor Obligations," 83 *Federal Register* 31923.

184 Stakeholders have disagreed with EPA's selection in other interstate air transport analyses, such as the 126(b) petitions regarding the 2008 and 2015 ozone NAAQS. See, for example, Delaware Department of Natural Resources and Environmental Control, *Testimony on EPA's Proposed Denial of Delaware's 126(b) Petitions*, p. 3.

185 The first factor stems from the D.C. Circuit's finding in *North Carolina v. EPA* that "emissions reductions required by the good neighbor provision should be evaluated considering the relevant attainment dates of downwind nonattainment areas impact by interstate transport." For the court case, see *North Carolina v. EPA,* 550 F.3d 896, 911- 12 (D.C. Cir. 2008).

186 For complete discussion about EPA's selection of the year 2023, see Page, October 27, 2017, pp. 4-7.

187 Page, October 27, 2017, pp. 4-5. See also EPA, *Ozone NAAQS Timelines*, https://www. epa.gov/ozone-pollution/ozone-naaqs-timelines.

188 EPA pointed to holdings from the U.S. Supreme Court and the D.C. Circuit Court "that the EPA may not require emissions reductions greater than necessary to achieve attainment and maintenance of the NAAQS in downwind areas." Page, October 27, 2017, pp. 4-6. See also *EPA v. EME Homer City Generation, L.P.*, 134 S. Ct. 1584, 1600-01 (2014); *EME Homer City Generation, L.P. v. EPA*, 795 F.3d 118, 127 (D.C. Cir. 2015).

189 Page, October 27, 2017, pp. 4-6.

EPA acknowledged that the year it chose—2023—is later than the attainment date for areas classified as "Serious" nonattainment (2008 ozone standard) but concluded that "it is unlikely that emissions control requirements could be promulgated and implemented by the serious area attainment date."[190]

The timing of EPA's proposed determination was driven in part by a court order. A federal district court in New York ordered EPA to propose determinations for five states by June 30, 2018, and finalize them by December 6, 2018.[191] EPA is under additional court-ordered and statutory deadlines to fully address the Good Neighbor provision with respect to the 2008 ozone standard. For example, another federal district court in California ordered EPA to address the Good Neighbor provision for Kentucky by June 30, 2018.[192] EPA is subject to statutory deadlines in 2018 and 2019 to address requirements for eight CSAPR Update states.[193]

Good Neighbor Determinations and the 2015 Ozone Standard

Evaluation of interstate ozone transport with respect to the 2015 ozone NAAQS is underway. EPA has conducted air quality modeling to inform the development and review of the Good Neighbor SIPs and issued the results in a memorandum in March 2018.[194] States have the option to use these modeling results—for example, projections of potential nonattainment and maintenance monitoring sites with respect to the 2015 ozone NAAQS in the year 2023—to develop their Good Neighbor SIPs. States are required to submit Good Neighbor SIPs with respect to the 2015 ozone standard to EPA by October 1, 2018.[195] EPA will then evaluate the adequacy of the SIPs and determine whether additional steps are necessary to address ozone transport.

EPA's March 2018 memorandum also identified "potential flexibilities" or "concepts" for developing the Good Neighbor SIPs, describing considerations for each step of the transport framework.[196] One of these considerations centered on international ozone

---

[190] Page, October 27, 2017, p. 6.

[191] The five states are Illinois, Michigan, Pennsylvania, Virginia, and West Virginia. Order, *New York v. Pruitt*, 1:18-cv-00406-JGK (S.D.N.Y. June 12, 2018).

[192] Order, *Sierra Club v. Pruitt*, No. 3:15-cv-04328 (N.D. Cal. May 23, 2017). EPA approved Kentucky's Good Neighbor SIP, concluding that Kentucky's SIP and Kentucky's CSAPR Update FIP "fully address the requirements of the good neighbor provision for the 2008 ozone NAAQS for Kentucky." EPA, "Air Plan Approval; Kentucky; 2008 Ozone NAAQS Interstate Transport SIP Requirements," 83 *Federal Register* 33730, July 17, 2018.

[193] The eight states are Indiana, Ohio, New Jersey, Maryland, Louisiana, Texas, Wisconsin, and New York. EPA, "Determination Regarding Good Neighbor Obligations," 83 *Federal Register* 31931.

[194] Memorandum from Peter Tsirigotis, Director, to Regional Air Division Directors, EPA Regions 1-10, March 27, 2018, https://www.epa.gov/sites/production/files/2018-03/docu ments/transport_memo_03_27_18_1.pdf.

[195] The CAA requires states to submit Good Neighbor SIPs to EPA within three years of a NAAQS revision. CAA §110(a)(1); 42 U.S.C. §7410(a)(1).

[196] EPA clarified that it is not recommending any changes and that any determinations about states' Good Neighbor obligations would undergo a notice-and-comment rulemaking. Tsirigotis, p. A-1.

contributions. Specifically, EPA seeks feedback on the evaluation of international ozone contributions when determining whether a state significantly contributes to or interferes with maintenance of a NAAQS. This "potential flexibility" might involve developing a "consensus on evaluation of the magnitude of international ozone contributions relative to domestic, anthropogenic ozone contributions" to nonattainment or maintenance receptors and consider whether to weigh the "air quality, cost, or emission reduction factors" differently in areas with relatively high contributions from international sources.[197] EPA also invited stakeholders to suggest additional concepts— "including potential EPA actions that could serve as a model"—for the way Good Neighbor obligations are translated to enforceable emissions limits.[198]

Issues for Congressional Consideration

SO_2 and NO_x emissions have declined in recent decades, with SO_2, annual NO_x, and ozone season NO_x emissions well below the 2016 CSAPR budgets (see *Figure 3* and *Figure 4*). EPA's analysis suggests that its regional SO_2 and NO_x programs have reduced interstate transport of PM2.5 and ozone in the eastern United States. EIA's national-scale analysis also points to a combination of broader market and regulatory factors contributing to emission reductions, in particular for SO_2.

Going forward, it is not clear whether emissions will remain well below CSAPR budgets given recent low allowance prices for ozone season NO_x and the supply of banked allowances that can be used in future years.[199] In addition, EPA has not yet issued a determination about whether ozone transport contributes to air quality problems with respect to the 2015 ozone standard.[200]

The agency has, therefore, not yet determined whether and how it will update the CSAPR budgets with respect to the 2015 ozone standard.

Stakeholder views on interstate air pollution transport vary, generally reflecting disagreements about the level of emissions that should be reduced and which sources—and states—bear responsibility for doing so. Some stakeholders have expressed concern that interstate transport continues to harm air quality.[201] For example, some stakeholders have expressed concern about transport of ozone and ozone precursor emissions to downwind

---

[197] Tsirigotis, p. A-3.

[198] Under CSAPR, EPA has specified emissions budgets in a FIP to be implemented through an air-quality assured trading program. EPA's March 2018 memo does not indicate whether or how that approach might change. Tsirigotis, p. A-3.

[199] For details about the number of banked allowances, see EPA, *2016 Program Progress*, pp. 44-50.

[200] EPA, "Determination Regarding Good Neighbor Obligations," 83 *Federal Register* 31923.

[201] In their recommendations to EPA regarding designation of nonattainment areas, Delaware, New Jersey, and Wisconsin attributed ozone violations in their jurisdictions to emissions transport from other states. See state recommendation letters to EPA, https://www.epa. gov/ozone-designations/2015-ozone-standards-state-recommendations-epa-responses-and-technical-support.

states—and the health impacts associated with ozone exposure—and stated that some coal-fired power plants do not make full use of "already-installed pollution controls" to reduce ozone precursor emissions.[202] As discussed earlier in this chapter, EPA has recently denied a 126(b) petition and proposed to deny others from states seeking additional upwind reductions in ozone precursors, in part because the agency disagreed with each state's technical analysis (see "Section 126(b) Petitions").[203] Among the stakeholders disagreeing with the agency's rejection of Connecticut's 126(b) petition was a regional organization that raised concern that EPA has not used existing CAA tools to "adequately address interstate ozone transport in a timely manner."[204] On the other hand, emissions are below CSAPR budgets, and other stakeholders have questioned the feasibility of additional reductions in ozone precursors. These stakeholders have raised concerns about the extent to which international or natural sources contribute to ambient ozone concentrations.[205] The following issues may inform deliberations about interstate air transport, particularly as EPA continues its assessment of Good Neighbor obligations with respect to the 2015 ozone standard.

NO_x Emission Trends

Major sources of NO_x emissions include power plants, industrial facilities, and mobile sources such as cars and trucks.[206] EPA reported that NO_x emissions are expected to decline in the future through a "combination of the implementation of existing local, state, and federal emissions reduction programs and changing market conditions for [power] generation

---

[202] U.S. Congress, House Committee on Science, Space, and Technology, Subcommittee on Environment, *State Perspectives on Regulating Background Ozone*, 115th Cong., 2nd sess., June 21, 2018. Testimony of Elena Craft, Environmental Defense Fund, pp. 1-5, 9, https://science.house.gov/sites/republicans.science.house.gov/files/documents/HHRG%E2%80%93115%E2%80%93SY18%E2%80%93WState-ECraft20180621_0.pdf.

[203] For example, Connecticut, Delaware, Maryland, and New York submitted 126(b) petitions related to compliance with the 2008 and/or 2015 ozone NAAQS. EPA, *Ozone National Ambient Air Quality Standards (NAAQS) Section 126 Petitions*, June 2018, https://www.epa.gov/ozone-pollution/ozone-national-ambient-air-quality-standards-naaqssection-126-petitions.

[204] Arthur N. Marin, Executive Director, Northeast States for Coordinated Air Use Management, letter to William Wehrum, Assistant Administrator, EPA Office of Air and Radiation, March 26, 2018, http://www.nescaum.org/documents/nescaum-comments-ct-126-petition-20180326.pdf/.

[205] U.S. Congress, House Committee on Science, Space, and Technology, Subcommittee on Environment, *State Perspectives on Regulating Background Ozone*, 115th Cong., 2nd sess., June 21, 2018. See testimony of Diane D. Rath, Alamo Area Council of Governments, pp. 5-7, https://science.house.gov/sites/republicans.science.house.gov/files/documents/HHRG %E2%80%93115%E2%80%93SY18%E2%80%93WState-DRath-20180621_0.pdf; and testimony of Timothy Franquist, Arizona Department of Air Quality, pp. 1-3, https://science.house.gov/sites/republicans.science.house.gov/files/documents/HHRG%E2%80%93115%E2%80%93SY18%E2%80%93WStateTFranquist-20180621_0.pdf. See also EPA, *Final Report on Review of Agency Actions that Potentially Burden the Safe, Efficient Development of Domestic Energy Resources Under Executive Order 13783*, October 25, 2017, p. 5, https://www.epa.gov/laws-regulations/final-report-review-agency-actions-potentially-burden-safe-efficient-development.

[206] EPA estimated that mobile sources accounted for over 50% of continental U.S. NO_x emissions. EPA implements separate programs to address mobile source emissions. EPA, "Determination Regarding Good Neighbor Obligations," 83 *Federal Register* 31918.

technologies and fuels."[207] EIA's projections, however, suggest that while coal-fired power generation declines in the reference scenario, power sector NO_x emissions remain relatively flat between 2017 and 2050, showing a total decline of 0.2%.[208] EPA noted that nonpower-sector sources may be "well-positioned to cost-effectively reduce NO_x" emissions compared to the power sector, but the agency also concluded that it has less certainty about nonpower-sector NO_x control strategies.[209]

The extent to which the current collection of federal and state programs—such as CSAPR and EPA mobile source programs that set tailpipe emission standards—improve air quality in areas not meeting the 2015 ozone standard is to be determined.[210] In 2015, EPA projected that existing rules (e.g., those addressing automobile emission and fuel economy standards and rules affecting power plants) would reduce ozone precursor emissions, regardless of whether EPA revised the ozone NAAQS.[211] EPA has subsequently proposed changes to some of these existing rules— specifically, greenhouse gas emission (GHG) standards for passenger cars and light trucks and existing coal-fired power plants.[212] In particular, the proposal for passenger cars and light trucks would freeze fuel economy and GHG standards at model year 2020 levels through model year 2026. The current GHG standards would decrease between model years 2020 and 2025 and were projected to decrease carbon dioxide as well as ozone precursor emissions.[213] In terms of power plants,

---

[207] EPA, "Determination Regarding Good Neighbor Obligations," 83 *Federal Register* 31926.

[208] EIA, *Annual Energy Outlook 2018*, Electricity Supply, Disposition, Prices, and Emissions, https://www.eia.gov/outlooks/aeo/data/browser/#/?id=8-AEO2018&region=0-.

[209] EPA, "Determination Regarding Good Neighbor Obligations," 83 *Federal Register* 31927.

[210] For more information about EPA's "Tier 3" requirements for tailpipe emissions, see CRS Report R43497, *Tier 3 Motor Vehicle Emission and Fuel Standards*, by Richard K. Lattanzio and James E. McCarthy.

[211] Specifically, EPA's 2015 analysis projected that the existing rules would bring monitored ozone levels to 70 ppb or lower in all but 14 counties (excluding California) of the 241 counties showing nonattainment with the 70 ppb level, based on 2012-2014 monitoring data. EPA's 2015 analysis also projected that the 2015 ozone revision would lead to further emission reductions. For discussion, see CRS Report R43092, *Implementing EPA's 2015 Ozone Air Quality Standards*, by James E. McCarthy and Kate C. Shouse. See also EPA, *Regulatory Impact Analysis of the Final Revisions to the National Ambient Air Quality Standards for Ground-Level Ozone*, September 2015, p. 26, https://www. regulations.gov/document?D=EPA-HQ-OAR-2013-0169-0057.

[212] On August 1, 2018, the National Highway Traffic Safety Administration and EPA proposed to amend certain existing Corporate Average Fuel Economy and GHG emission standards for passenger cars and light trucks. The proposal has not yet been published in the *Federal Register*. A pre-publication version can be viewed at https://www.epa.gov/ sites/production/files/2018-08/documents/safe-my-2021-2026-cafe-ld-ghg-nhtsa-epa-nprm-2018- 08-02.pdf. On August 20, 2018, EPA proposed, among other things, to replace the Clean Power Plan, an Obama Administration rule that would limit carbon dioxide, SO_2, and NO_x emissions from existing fossil-fuel-fired power plants. The proposed rule has not yet been published in the *Federal Register* but can be viewed on EPA's website. See EPA, "Emission Guidelines for Greenhouse Gas Emissions from Existing Electric Utility Generating Units; Revisions to Emission Guideline Implementing Regulations; Revisions to New Source Review Program," https://www.epa.gov/stationary-sources-air-pollution/ proposal-affordable-clean-energy-ace-rule.

[213] For summary of the current GHG standards for cars and light trucks (model years 2017 to 2025) and projected emission impacts, see EPA, *EPA and NHTSA Set Standards to Reduce Greenhouse Gases and Improve Fuel Economy for Model Years 2017-2025 Cars and Light Trucks*, August 2012, https://nepis.epa.gov/Exe/ZyPDF.cgi/P100EZ7C.PDF?Dockey =P100EZ7C.PDF. For additional discussion about these standards, see EPA, *Regulatory Impact Analysis: Final Rulemaking for 2017-2025 Light-Duty Vehicle Greenhouse Gas Emission Standards and Corporate Average Fuel Economy Standards*, August 2012, https://nepis.epa.gov/Exe/ZyPDF.cgi/P100EZI1.PDF? Dockey=P100EZI1.PDF. See also CRS Report R45204, *Vehicle Fuel Economy and Greenhouse Gas Standards: Frequently Asked Questions*, by Richard K. Lattanzio, Linda Tsang, and Bill Canis.

EPA concluded that its Affordable Clean Energy proposal to replace the Clean Power Plan would increase carbon dioxide, SO_2, and NO_x emissions from the power sector relative to a scenario with implementation of the Clean Power Plan.[214] While the agency has not yet finalized these changes, they may have implications for levels of ozone precursor emissions. That is, regulatory changes affecting emissions in one sector—such as automobiles—may affect ozone NAAQS implementation as states seek to ensure the necessary emission reductions are achieved across all sources—mobile and stationary—in the state.

Incentives for NO_x Reductions

A recent market report concluded that current NO_x allowance prices—which are lower than the marginal cost of NO_x reductions—may ultimately lead to higher emissions.[215] While EPA has set state-specific emission budgets for CSAPR states intended to address interstate ozone transport with respect to the 2008 ozone standard, it is not clear whether these budgets will be sufficient to address Good Neighbor obligations under the more stringent 2015 ozone standard.

In light of this trend in NO_x allowance prices, some have questioned whether additional regulatory incentives may be necessary for states to fulfill Good Neighbor obligations.[216] Some states have urged EPA to implement additional regulatory requirements through 126(b) petitions.[217] For example, Delaware's 126(b) submission to EPA concluded that "[a]dditional regulatory incentive is required to ensure that the existing [Electric Generating Unit] NO_x controls are consistently operated in accordance with good pollution control practices."[218]

Related EPA Air Quality Initiatives

Current Trump Administration air quality initiatives may indirectly affect consideration of states' Good Neighbor obligations. The Administration has established a

---

[214] EPA, Regulatory Impact Analysis for the Proposed Emission Guidelines for Greenhouse Gas Emissions from Existing Electric Utility Generating Units; Revisions to Emission Guideline Implementing Regulations; Revisions to New Source Review Program, August 2018, pp. ES-8 to ES-9, https://www.epa.gov/sites/production/files/2018-08/documents/utilities_ ria_proposed_ace_2018-08.pdf.

[215] Evolution Markets, *Market Update*, p. 2.

[216] Thomas F. McNevin, "Recent Increases in Nitrogen Oxide (NO_x) Emissions from Coal-Fired Electric Generating Units Equipped with Selective Catalytic Reduction," *Journal of the Air and Waste Management Association*, vol. 66, no. 1 (2016), pp. 66-75.

[217] EPA, *Ozone National Ambient Air Quality Standards (NAAQS) Section 126 Petitions*, June 2018, https://www.epa.gov/ozone-pollution/ozone-national-ambient-air-quality-standards-naaqs-section-126-petitions.

[218] David Small, Secretary, Delaware Department of Natural Resources and Environmental Control, letter to Gina McCarthy, EPA Administrator, November 28, 2016, p.29, http://www.dnrec.delaware.gov/Air/Documents/126b%20Petitions/Conemaugh%20126%20Petition.pdf.

"NAAQS Reform" initiative that, among other things, seeks to streamline the NAAQS review process and obtain Clean Air Scientific Advisory Committee advice regarding background pollution and potential adverse effects from NAAQS compliance strategies.[219] EPA has also created an Ozone Cooperative Compliance Task Force in response to some stakeholders' concerns about international and long-range ozone transport as well as monitoring and modeling issues.[220] Limited information is available about the Ozone Cooperative Compliance Task Force and what actions it may undertake.

In March 2018, EPA reiterated its interest in these particular ozone issues when it published air quality projections meant to inform Good Neighbor evaluations with respect to the 2015 ozone standard. Specifically, EPA's memorandum sought comment on "potential flexibilities" for developing the Good Neighbor SIPs, describing considerations for each step of the transport framework, including assessment of international ozone transport.[221]

---

[219] E. Scott Pruitt, EPA Administrator, letter to EPA Assistant Administrators, May 9, 2018, https://www.epa.gov/sites/production/files/2018-05/documents/image2018-05-09-173219.pdf. See also EPA, *Final Report on Review of Agency Actions That Potentially Burden the Safe, Efficient Development of Domestic Energy Resources Under Executive Order 13783*, October 25, 2017, p. 3, https://www.epa.gov/laws-regulations/final-report-review-agency-actions-potentiallyburden-safe efficient-development.

[220] EPA, *Final Report*, p. 3.

[221] EPA clarified that it is not recommending any changes and that any determinations about states' Good Neighbor obligations would undergo a notice-and-comment rulemaking. Tsirigotis, p. A-1.

In: The Clean Air Act and The Clean Power Plan
Editor: Claire Ferguson

ISBN: 978-1-53614-825-1

Chapter 6

VOLKSWAGEN, DEFEAT DEVICES, AND THE CLEAN AIR ACT: FREQUENTLY ASKED QUESTIONS*

Bill Canis, Richard K. Lattanzio, Adam Vann and Brent D. Yacobucci

SUMMARY

The German automotive manufacturer Volkswagen Automotive Group (VW) has admitted to installing a software algorithm in several of its diesel-fueled vehicle engines that acts as a "defeat device": the software detects when the vehicle is undergoing compliance testing and activates certain pollution control devices to reduce tailpipe emissions. During normal driving situations, however, the control devices are turned off, resulting in higher emissions of nitrogen oxide (NO_x) and other air pollutants than claimed by the company. Federal and California regulators and the European Union (EU) have examined the use of this software, which was reportedly installed in 11 million vehicles worldwide.

A summary of federal and state actions includes the following:

- September 18, 2015: the U.S. Environmental Protection Agency (EPA) issued a notice of violation (NOV) of the Clean Air Act (CAA) to VW, contending that 2.0 liter Volkswagen and Audi diesel cars (model years 2009-2015) include software that circumvents EPA standards for NO_x, allowing emissions up to 40 times the standard.

* This is an edited, reformatted and augmented version of a Congressional Research Service report, Publication No. R44372, dated September 1, 2016.

- November 2, 2015: EPA issued a second NOV alleging that VW installed defeat devices in light-duty diesel vehicles equipped with 3.0 liter engines for model years 2014-2016, resulting in NO_x emissions increases nine times the EPA standard.
- January 4, 2016: the U.S Department of Justice (DOJ) filed a civil complaint against VW on behalf of EPA in federal court alleging that nearly 600,000 diesel vehicles had illegal defeat devices installed, thereby impairing emissions controls and causing harmful air pollution in excess of EPA standards.
- EPA stated that it will not grant a certificate of conformity for VW's model year 2016 diesel vehicles, thus halting sales of these vehicles in the United States.
- The California Air Resources Board initiated an investigation into VW's use of this "defeat device," and, on January 12, 2016, issued a NOV to VW, alleging that "approximately 75,688 California vehicles do not conform to State law."
- June 28, 2016: EPA, the state of California and the Federal Trade Commission (FTC) announced a settlement with VW with regard to its 2.0 liter vehicles, including a $10 billion buyback of affected cars from consumers, and $4.7 billion to mitigate pollution and support zero emission vehicle technology.

This chapter is organized as a series of frequently asked questions. It focuses on a description of modern diesel technologies, their market and emissions profiles, and some potential reasons that could underlie the use of defeat devices. It summarizes the specific allegations filed against VW under the CAA, the current status of federal and state investigations, and the civil and potential criminal penalties that may result. Further, the report introduces several outstanding issues currently under debate, including whether EPA has sufficient resources to monitor vehicle emissions, whether the current penalty structure is sufficient, why EPA failed to detect VW's defeat device when there have been similar cases in the past, and whether VW's response to the emissions problem and efforts to provide restitution to U.S. customers have been adequate.

INTRODUCTION

The German automotive manufacturer Volkswagen Automotive Group (VW)[1] has admitted to installing a software algorithm in several of its diesel-fueled vehicle engines that acts as a "defeat device": the software detects when the vehicle is undergoing official compliance testing and activates certain pollution control devices to reduce tailpipe emissions. During normal driving situations, however, the control devices are turned off, resulting in higher emissions of nitrogen oxides (NO_x) and other air pollutants than claimed by the company. Such software could allow higher on-road performance and fuel economy than otherwise attainable with fully active emissions systems. Federal and California regulators examined the use of this software— reportedly installed in 11 million vehicles worldwide from model years (MY) 2009 to 2016—and announced a $14.7 billion partial

[1] The allegations discussed in this report were against a number of related entities, including Volkswagen AG; Audi AG; Volkswagen Group of America, Inc.; Volkswagen Group of America Chattanooga Operations, LLC; Dr. Ing H.c. F. Porsche AG; and Porsche Cars North America, Inc. The report refers to these related entities collectively as "Volkswagen Automotive Group" or "VW."

settlement in June 2016. The European Union (EU) is also examining the use of the software.

ALLEGATIONS AGAINST VOLKSWAGEN

What Actions Has the U.S. Government Taken Against VW?

On September 18, 2015, the U.S. Environmental Protection Agency (EPA) issued a notice of violation (NOV)[2] to VW, contending that Volkswagen and Audi vehicles with 2.0 liter diesel engines (MY2009-MY2015) include software that circumvents EPA emissions standards for nitrogen oxide (NO_x) emissions. EPA stated that when the emissions equipment is disabled, NO_x emissions are up to 40 times greater than the standard. The California Air Resources Board (CARB) also initiated an investigation into VW's use of this "defeat device."[3] These allegations cover roughly 499,000 diesel passenger cars sold in the United States.

Following the September 18 notice to VW, EPA initiated testing of all U.S. 2015 and 2016 light-duty diesel models to detect potential defeat devices. On November 2, 2015, EPA issued a second NOV alleging that VW installed defeat devices in light-duty diesel vehicles equipped with 3.0 liter engines for MY2014-MY2016, resulting in NO_x emissions nine times the EPA standard. This notice affects 85,000 vehicles sold since MY2009.[4]

The U.S. Department of Justice (DOJ) filed a civil complaint on behalf of EPA in federal court on January 4, 2016.[5] DOJ alleged that nearly 600,000 diesel vehicles had illegal defeat devices installed, thereby impairing emissions controls and causing harmful air pollution in excess of EPA standards. The complaint also alleged that VW violated the Clean Air Act (CAA)[6] by selling vehicles that are designed differently from what it stated in applications for certification to EPA and CARB. DOJ sought injunctive relief and the assessment of unspecified civil penalties; other legal remedies may be pursued as well,

---

[2] EPA, "EPA, California Notify Volkswagen of Clean Air Act Violations," press release, September 18, 2015, http://yosemite.epa.gov/opa/admpress.nsf/a883dc3da7094f97852572a00065d7d8/dfc8e33b5ab162b985257ec40057813b!OpenDocument.

[3] CARB, "ARB Letter to VW," September 18, 2015, http://arb.ca.gov/newsrel/in_use_ compliance_letter.htm.

[4] EPA, "EPA, California Notify Volkswagen of Additional Clean Air Act Violations," press release, November 2, 2015, http://yosemite.epa.gov/opa/admpress.nsf/0/4A45A5661216E66 C85257EF10061867B, and EPA, "United States Files Complaint Against Volkswagen, Audi and Porsche for Alleged Clean Air Act Violations," press release, January 4, 2016, http://yosemite.epa.gov/opa/admpress.nsf/bd4379a92ceceeac8525735900400c27/ac7b52362207dad785257f300060442e!OpenDocument.

[5] DOJ, "United States Files Complaint Against Volkswagen, Audi and Porsche for Alleged Clean Air Act Violations," press release, January 4, 2016, http://www.justice.gov/opa/pr/united-states-files-complaint-against-volkswagen-audiand-porsche-alleged-clean-air-act.

[6] 42 U.S.C. §§7401-7671q.

such as criminal charges. Parties have settled regarding some of these legal remedies, while others remain unresolved.

How Were the Alleged Emissions Violations Brought to the Attention of EPA?

EPA and CARB were alerted to the emissions violations by researchers at West Virginia University (WVU) working under a contract with the International Council on Clean Transportation (ICCT), a nonprofit environmental research organization. As part of a study of on-road emissions from diesel vehicles, the WVU researchers found emissions levels for some vehicles far exceeded U.S. certification standards. The study was part of a larger investigation by ICCT motivated by reports that some European-made diesel vehicles had passed emissions tests but had much higher real-world NO_x emissions. (EU emissions standards apply only at the time a vehicle is produced; surveillance testing, mandatory emissions system warranties, and other features of U.S. rules are not incorporated in EU regulations.)

According to the EPA notice, VW initially indicated that the excess emissions resulted from a software problem that could be addressed by a voluntary recall. Ultimately, EPA found that software installed in the vehicles' computers sensed when the vehicles were being tested and activated a lower-emissions mode. Thus, nonstandard testing was necessary to reveal VW's actions. Such software could circumvent the "diesel dilemma," discussed below, and allow higher on-road performance and fuel economy than otherwise attainable with fully active emissions systems. WVU's testing indicated a BMW diesel vehicle was able to meet emissions targets. Thus, emissions compliance does not appear to be one of technical feasibility. It should also be noted that while ICCT has found other diesel vehicles that exceed European or U.S. NO_x standards in real-world use, VW is so far the only automobile manufacturer accused of using defeat devices.

What Is the Technology at Issue? What Is a TDI Diesel?

Diesel engines are internal combustion engines that use heat generated by fuel compression to ignite the diesel fuel.[7] Gasoline-powered engines use spark plugs and other components to ignite the fuel, fire the pistons, and drive the car. Otherwise, gasoline and diesel engines are similar. Diesel fuel—which is of a different chemical composition and contains more energy per unit of volume than gasoline—combined with compression ignition[8] is potentially a more energy-efficient process and in general delivers more power than gasoline. The 2015 EPA *Fuel Economy Guide* notes that

---

[7] The process is called compression ignition. Andrew Norman, John Corinchock, and Robert Scharff, *Diesel Technology* (Tinley Park, Illinois: The Goodheart-Willcox Company, 1998), p. 3.

[8] It has been estimated that diesel engines convert about 40% of their fuel energy into useable vehicle energy, compared with only 25% fuel efficiency in gasoline vehicles. Ibid., p. 13.

> Diesel engines are inherently more energy-efficient, and diesel fuel contains roughly 10%–15% more energy per gallon than gasoline. In addition, new advances in diesel engine technology have improved performance, reduced engine noise and fuel odor, and decreased emissions of harmful air pollutants. Ultra-low sulfur diesel fuels also help reduce emissions from these vehicles.[9]

Diesel engines generally last longer than gasoline engines and retain a higher resale value. However, diesel engines are more expensive to manufacture than gasoline engines and retail for more.[10] They generally emit greater quantities of NO_x and particulate matter (PM) that require pollution control devices not found on gasoline vehicles.

Diesel engine technology has changed in recent years as automakers have sought to find new ways to raise fuel economy and reduce emissions so they can meet new federal (and EU) greenhouse gas standards. Providing direct injection of fuel into the engine combustion chamber and turbocharging the air used to burn the fuel[11] are two ways in which the goals of higher fuel efficiency and lower emissions can be met. VW's technology is called Turbocharged Direct Injection (TDI).[12] Similar technology is found on other diesel-fueled passenger cars.[13]

What VW Vehicles Are Affected?

The motor vehicles that had defeat devices installed were all diesels manufactured in Europe or the United States. It has been estimated that about 584,000 diesel passenger cars sold in the United States since MY2009 are equipped with a defeat device. See *Table 1* for affected vehicles. EPA did not grant certificates of conformity for VW's MY2016 diesel vehicles, thus halting sales of these vehicles in the United States.[14]

What Is VW's U.S. Market?

Volkswagen is a German company established in 1933, with manufacturing operations around the world. In 2015, VW was the world's second-largest automaker after Toyota. In

---

[9] Diesel vehicles' fuel economy is seen in EPA's comparative vehicle ratings: a 2.0 liter, model year (MY) 2015 diesel Jetta gets 31 miles per gallon (mpg) in the city and 45 mpg on the highway, for a combined 36 mpg. A 2.0 liter gasoline-powered Jetta gets 24 mpg in the city and 32 on the highway, for a combined 27 mpg. EPA and Department of Energy, *Model Year 2015 Fuel Economy Guide*, November 19, 2015, pp. 13 and 29; https://www.fueleconomy.gov/ feg/pdfs/guides/FEG2015.pdf.

[10] A 2015 VW Jetta's MSRP (manufacturer's suggested retail price) was $20,760; the MSRP for the TDI diesel version of the Jetta was $26,665; http://www.edmunds.com and http://www.edmunds.com/volkswagen/ jetta/2015/diesel.

[11] A turbocharger increases the amount of air supplied to the engine at higher than normal pressure, using a turbine powered by exhaust gases. This process improves both combustion efficiency and power output. Isuzu Motors Limited, *Technology for Cleaner Diesel: CO_2 Reduction Technologies*, http://www.isuzu.co.jp/world/ technology/clean/ cleaner01.html.

[12] Volkswagen, *Technical Glossary*, TDI, http://en.volkswagen.com/en/innovation-and-technology/technical-glossary/ tdi.html.

[13] The diesel-fueled Chevrolet Cruze sedan's technology is called Clean Turbo Diesel; http://www.GM.com.

[14] As reported in the September 18, 2015, NOV.

"EPA would work with the manufacturer to fix them, either through voluntary manufacturer action or, if necessary, through an ordered emissions recall."[25]

In addition to the manufacturer-controlled IUVP, EPA also conducts limited "surveillance testing" at its laboratory in Ann Arbor, MI. EPA selects vehicles for such testing from IUVP data, EPA certification data, consumer complaints, and random selection. Each year EPA tests a few dozen vehicles. In the wake of the VW allegations, EPA issued guidance to manufacturers on September 25, 2015, that it may require additional testing to investigate potential defeat devices.[26]

Why Might Have VW Installed a Defeat Device?

VW has not stated why the defeat devices were installed. Experts in automotive technology have said that disengaging the pollution controls on a diesel-fueled car can yield better performance, including increased torque and acceleration.[27] Further, several types of emissions control technologies require fuel to run; thus, disengaging them would return better fuel economy for the vehicle. In the case of VW, such modifications may have been intended to allow vehicles designed for the European Union market to meet more stringent U.S. NO_x regulations, avoiding additional investment for the comparatively small U.S. diesel vehicle market.

Consumer Reports concluded that VW may have used the defeat devices to increase fuel economy and vehicle performance. It tested MY2011 and MY2015 VW TDI diesel vehicles with and without the defeat device engaged and found a "noticeable decline in fuel economy for both models" when the defeat device was not engaged.[28]

THE INVESTIGATION AND POTENTIAL OUTCOMES

What Violations of the CAA Were Alleged in DOJ's Civil Complaint?

On January 4, 2016, DOJ filed a civil complaint against VW in the U.S. District Court for the Eastern District of Michigan[29] based on the allegations regarding installation of

[25] See EPA, "2008 Progress Report: Vehicle and Engine Compliance Activities," EPA-420-R-10-022, August 2010, http://www3.epa.gov/otaq/about/420r10022.pdf#page=31.

[26] EPA, "EPA Update on Recent Volkswagen Announcement," press release, September 18, 2015, http://yosemite.epa.gov/opa/admpress.nsf/6424ac1caa800aab85257359003f5337/6579a74e2ed0039185257ecb004f34cf!OpenDocument.

[27] See Coral Davenport and Jack Ewing, "VW Is Said to Cheat on Diesel Emissions; U.S. to Order Big Recall," *New York Times*, September 18, 2015.

[28] Jeff Bartlett, "Guide to the Volkswagen Emissions Recall," *Consumer Reports*, January 4, 2016, http://www.consumerreports.org/cro/cars/guide-to-the-volkswagen-dieselgate-emis sions-recall .

[29] The matter was subsequently transferred to ongoing multi-district litigation before the Northern District of California, as discussed below.

defeat devices in 2.0L diesel vehicles, as described in this chapter.[30] The complaint made four claims for relief:

- that Volkswagen sold, offered for sale, introduced into commerce, delivered for introduction into commerce, or imported vehicles that did not conform in all material respects with the specifications in the COCs purported to cover them, in violation of Section 203(a)(1) of the CAA;[31]
- that Volkswagen manufactured, sold, offered for sale, or installed parts or components in certain vehicles intended for use with motor vehicles where a principal effect of the part or component is to bypass, defeat, or render inoperative a device or element of design installed in compliance with CAA regulations, in violation of Section 203(a)(3)(B) of the CAA;[32]
- that certain auxiliary emission control devices installed by Volkswagen had the effect of removing or rendering inoperative devices or elements of the emissions control system installed in new vehicles in compliance with CAA regulations, in violation of Section 203(a)(3)(A) of the CAA; and
- that Volkswagen failed to disclose the existence of the auxiliary emission control devices in the COC applications for test groups for new vehicles, in violation of the reporting requirements found in Section 203(a)(2) of the CAA.[33]

What Legal Consequences Might Have VW Faced for the Violations Alleged in the January 2016 DOJ Civil Complaint of the CAA?

Part A of Title II of the CAA, which deals with emissions standards for moving sources, does not provide for criminal penalties. It is possible, however, that VW or its officials could face criminal charges based on other statutes.[34] For example, DOJ could pursue charges under federal mail fraud or wire fraud prohibitions if VW has used either medium to convey false information in service of a "scheme or artifice to defraud."[35]

In addition, CAA violations like those described in the January 2016 Complaint can trigger civil penalties. Section 205 of the CAA sets forth civil penalties for these violations.

---

[30] *U.S. v. Volkswagen AG et al.*, 2:16-cv-10006 (Complaint), January 4, 2016. The complaint is available at http://www.justice.gov/opa/file/809826/download.

[31] Complaint, ¶¶ 103-105, citing 42 U.S.C. §7522(a)(1).

[32] Complaint, ¶¶ 109-111, citing 42 U.S.C. §7522(a)(3)(B).

[33] Complaint, ¶¶ 123-129, citing 42 U.S.C. §7522(a)(2).

[34] Reports emerged in August 2016 suggesting that DOJ is investigating the possibility of criminal liability for Volkswagen for the actions described here, although the reports did not mention a specific criminal charge. See, for example, Mike Spector and Aruna Viswanatha, "U.S. Said to Uncover Evidence of Criminal Acts in VW Probe," *Wall Street Journal*, August 15, 2016.

[35] 18 U.S.C. §§1341, 1343.

The January 4, 2016, complaint provided further details regarding the potential penalties applicable to Volkswagen:

- For violations of the COC requirements found in Section 203(a)(1) of the CAA, the complaint stated that failure to comply with the requirements is a separate offense for each motor vehicle, and that pursuant to Section 205(a) of the CAA, VW could be liable for civil penalties of up to $32,500 per vehicle for each violation occurring before January 13, 2009, and for civil penalties of up to $37,500 for each violation occurring on or after January 13, 2009.[36]
- For violations of the prohibition on installation of a "defeat device" found in Section 203(a)(3)(B) of the CAA, the complaint again stated that failure to comply with the requirements is a separate offense for each motor vehicle, and that pursuant to Section 205(a) of the CAA, VW could be liable for civil penalties of up to $2,750 per part or component installed vehicles prior to January 13, 2009, and for civil penalties of up to $3,750 per part or component installed on or after January 13, 2009.[37]
- For violations of the prohibition on tampering found in Section 203(a)(3)(A) of the CAA, the complaint stated that each vehicle equipped with an auxiliary emission control device that removed or rendered inoperative devices or elements of the emissions control system installed in new vehicles constitutes a separate violation, and that pursuant to Section 205(a) of the CAA, VW could be liable for civil penalties of up to $32,500 per vehicle for each violation occurring before January 13, 2009, and for civil penalties of up to $37,500 for each violation occurring on or after January 13, 2009.[38]
- For violations of the reporting requirements of Section 203(a)(2) of the CAA, the complaint stated that each failure to provide reports or information constitutes a separate violation, and that pursuant to Section 205(a) of the CAA, VW could be liable for civil penalties of up to $32,500 per day of violation occurring before January 13, 2009 and up to $37,500 per day of violation occurring on or after January 13, 2009.[39]
- VW also faces the possibility of injunctive relief for each of the alleged violations pursuant to Section 204(a) of the CAA.[40] This allows the court to take action to restrain VW from continued violations of the listed provisions.

---

[36] Complaint ¶¶ 106-107, citing 42 U.S.C. §7524(a). The shift in the penalty cap on January 13, 2009, is a product of a regulatory adjustment to civil monetary penalties by EPA on that date pursuant to the Federal Civil Penalties Act of 1990 (28 U.S.C. §2461 note) and the Debt Collection Improvement Act of 1996 (31 U.S.C. §3701 note).

[37] Complaint, ¶¶ 112-113, citing 42 U.S.C. §7524(a).

[38] Complaint, ¶¶ 119-121, citing 42 U.S.C. §7524(a)

[39] Complaint, ¶¶ 129-131, citing 42 U.S.C. §7524(a).

[40] Complaint, ¶¶ 107, 113, 121 and 131, citing 42 U.S.C. §7523(a).

It should be noted that the potential penalties outlined above are not generally imposed, as automakers charged with such violations generally negotiate a lower penalty with EPA to settle the case. For example, DOJ sued Toyota for $58 billion in environmental violations more than a decade ago, but Toyota settled with the government, resulting in, among other things, a $34 million penalty.[41] This appears to have been the avenue chosen by VW and DOJ in this instance, as discussed below.

In addition, it should be noted that Section 205 of the Clean Air Act specifically requires courts to consider a violator's ability to pay a penalty and remain in business when assessing a civil penalty. Section 205 states the following:

> In determining the amount of any civil penalty to be assessed under this subsection, the court shall take into account the gravity of the violation, the economic benefit or savings (if any) resulting from the violation, the size of the violator's business, the violator's history of compliance with this title, action taken to remedy the violation, the effect of the penalty on the violator's ability to continue in business, and such other matters as justice may require.[42]

What Are VW's Obligations Under the Proposed DOJ Civil Settlement?

On January 15, 2016, DOJ's litigation, which had been initiated in the Eastern District of Michigan, was transferred to the U.S. District Court for the Northern District of California in order to consolidate it with an ongoing litigation in which a number of private parties have filed claims based on VW's alleged wrongdoing.[43] Subsequently, VW reached a series of settlements that may resolve many of its liability issues related to the installation of defeat devices in 2.0L diesel vehicles.[44] In addition to filed settlements related to alleged violations of Federal Trade Commission (FTC) regulations[45] and private liability claims,[46] a Proposed Partial Consent Decree filed on June 28, 2016, appears to resolve VW's potential liability for the violations of CAA requirements described above.[47] The consent

---

[41] Julia Edwards and Georgina Prodhan, "Volkswagen faces billions in fines as U.S. sues for environmental violations," January 6, 2016 and U.S. v Toyota Motor Corp et al., Civ. No. 99-1888, Consent Decree (D.C. Cir 2003); http://www.epa.gov/sites/production/ files/documents/toyotacd.pdf.

[42] 42 U.S.C. §7524(b).

[43] *U.S. v. Volkswagen AG*, 2:16-cv-10006, Notice of Transfer. (January 15, 2016).

[44] Note that Volkswagen also appears to have reached settlement on claims brought by the Federal Trade Commission (FTC) and private parties.

[45] *In re: Volkswagen "Clean Diesel" Marketing, Sales Practices and Products Liability Litigation*; MDL No. 2672-CRB, [Proposed] Partial Stipulated Order for Permanent Injunction and Monetary Judgment (June 28, 2016).

[46] *In re: Volkswagen "Clean Diesel" Marketing, Sales Practices and Products Liability Litigation*; MDL No. 2672-CRB, Consumer Class Action Settlement Agreement and Release (Amended) (July 26, 2016).

[47] In re: Volkswagen "Clean Diesel" Marketing, Sales Practices and Products Liability Litigation; MDL No. 2672-CRB, Partial Consent Decree (June 28, 2016).

decree also appears to resolve VW's potential liability for violations of the California Health and Safety Code and the California Code of Regulations.[48]

According to the Notice of Lodging of Proposed Partial Consent Decree issued by DOJ on July 6, 2016,[49] "[t]he three settlements resolve separate claims but offer coordinated relief."[50] This coordinated relief provided for in the three settlements includes the following:

- VW must offer all eligible owners and lessees of eligible vehicles the option to have VW buy back their cars or to terminate their leases at no cost. VW may submit for EPA and CARB review and approval a proposal for modifying these 2.0 liter vehicles to reduce emissions. If EPA and CARB approve an emissions modification for any category of the 2.0 liter vehicles, VW may offer all eligible owners and lessees the additional option of receiving an emissions modification in lieu of a buyback.
- VW must fund a trust over three years in the total amount of $2.7 billion, which states, Puerto Rico, the District of Columbia, and Indian tribes can use to perform specified NO_x mitigation projects.
- VW must achieve a recall rate (through the buyback, lease termination, scrapped vehicles, and the emissions modification option, if approved) of 85% by June 30, 2019. If it fails to do so, VW must augment the mitigation trust fund discussed below by $85 million for each 1% that it falls short of the 85% rate. VW must also achieve a separate 85% recall rate for vehicles in California, and must pay $13.5 million to the mitigation trust (solely for mitigation projects in California) for each 1% that it falls short of this target.
- In connection with the buyback, VW must pay eligible owners no less than the cost of the retail purchase of a comparable replacement vehicle of similar value, condition, and mileage as of September 17, 2015, the day before the existence of the defeat devices was made known to the public.[51] The Decree acknowledges that VW may satisfy this obligation through offering the payments required by the FTC Order and the Class Action Settlement, which are at least equal to the retail

---

[48] The California Air Resources Board (CARB), which collaborated with EPA on the investigation of VW, issued a Notice of Violation (CANOV) to VW on January 12, 2016.[48] In the notice, CARB alleged 13 different violations of the California Health and Safety Code or the California Code of Regulations triggered by the sale or offering for sale of approximately 75,688 vehicles in the state of California. The alleged violations range from failure to disclose all auxiliary emission control devices in its applications for CARB certification to violation of emission warranty requirements. The California Notice of Violation can be viewed at http://arb.ca.gov/msprog/vw_info/nov_vw.pdf.

[49] Notice of Lodging of Proposed Partial Consent Decree Under the Clean Air Act, 88 Fed. Reg. 44,501 (July 6, 2016).

[50] *Id.*

[51] This is the "retail replacement value."

replacement value. The buyback/lease termination program under the Decree remains open for two years after the Decree is entered by the court. (See Decree Section IV.A and Appendix A.) If EPA and CARB approve an emissions modification, VW must offer it to consumers indefinitely.

- VW must invest $2 billion over a 10-year period to support the increased use of zero emission vehicle (ZEV) technology[52] in the United States, including the development and maintenance of ZEV charging stations and infrastructure.[53]

Note that the proposed consent decree, the FTC settlement, or the private settlement would not absolve VW of potential civil penalties claimed by the federal government in the January 2016 Complaint,[54] or of any potential criminal charges related to this matter.

What Is the Next Step in the Federal Proceedings?

The publication of notice of the Proposed Consent Decree triggered a 30-day public comment period, which concluded on August 6, 2016. If the Proposed Consent Decree is approved and adopted by the U.S. District Court for the Northern District of California following review of the public comments, it will become binding on the parties, and DOJ's claims for injunctive relief will be dismissed. It is not clear how the outstanding DOJ claims for civil penalties will be addressed going forward.

Have Similar Violations Happened Before?

Since the 1970s, EPA has repeatedly found manufacturers using defeat devices in violation of the CAA. When it determines that defeat devices have been installed, EPA begins enforcement proceedings. In response, automakers often voluntarily recall the vehicles and/or settle with EPA and DOJ. For example, in 1998 Honda and Ford agreed to pay $267 million and $7.8 million, respectively, for fines and pollution mitigation. Other cases where EPA has accused manufacturers of installing defeat devices include automakers VW (1973), Chrysler (1973), and General Motors (1995); heavy-duty engine manufacturers Caterpillar, Cummins, Detroit Diesel, Mack, Navistar, Renault and Volvo (1998); parts manufacturers Casper's Electronics (2013); Edge Products (2013); and Harley-Davidson (2016). For a selected list of cases involving defeat devices, see Table 2.

---

[52] ZEV technology measures pollution at a vehicle's tailpipe; battery electric vehicles, plug-in hybrid electric vehicles, and fuel cell vehicles are examples of ZEVs.

[53] Notice of Lodging of Proposed Partial Consent Decree Under the Clean Air Act, 88 Fed. Reg. 44,501 (July 6, 2016).

[54] *Id.*

Table 2. Defeat Device Cases under Title II of the Clean Air Act

Company	Year	Civil Penalties	Additional Relief	Affected Vehicles/Equipment	Link
Harley-Davidson	2016	$12 million	Mandatory recall/repurchase program; destruction of recalled devices; $3 million for cleaner burning wood stoves to mitigate air pollution	340,000 "super tuner" aftermarket defeat devices added to motorcycles, and 12,000 MY2006-MY2008 motorcycles not covered by an EPA certificate of conformity.	https://www.epa.gov/enforcement/harley-davidson-clean-air-act-settlement
Volkswagen	2015	In process		584,000 MY2009-MY2015 diesel cars	http://www3.epa.gov/otaq/cert/violations.htm
Casper's Electronics	2013	$80,000	Mandatory recall/repurchase program; destruction of recalled devices	44,000 aftermarket "oxygen sensor simulators"	http://www2.epa.gov/enforcement/caspers-electronics-inc-clean-air-act
Edge Products LLC	2013	$500,000	Mandatory repurchase program; $157,600 in rebates to upgrade older wood-burning stoves	9,000 aftermarket electronic devices sold from 2009 to 2011 that allowed for the removal of emissions equipment	http://www2.epa.gov/enforcement/edge-products-llc-settlement
Caterpillar, Inc.	1998	$25 million	$35 million for environmental projects[a]	320,000 heavy-duty diesel engines produced from 1988 to 1998	http://www2.epa.gov/enforcement/caterpillar-inc-diesel-engines-settlement
Cummins Engine Company	1998	$25 million	$35 million for environmental projects[a]	400,000 heavy-duty diesel engines produced from 1991 to 1998	http://www2.epa.gov/enforcement/cummins-engine-company-diesel-engine-clean-air-act-settlement
Detroit Diesel Corp.	1998	$12.5 million	$12 million for environmental projectsa	430,000 heavy-duty diesel engines produced from 1988 to 1998	http://www2.epa.gov/sites/production/files/documents/detroit-cd.pdf
Mack Trucks, Inc. and Renault Vehicles Industrials, U.S.A.	1998	$13 million	$18 million for environmental projects[a]	90,000 heavy-duty diesel engines produced from 1990 to 1998	http://www2.epa.gov/enforcement/mack-trucks-diesel-engine-settlement
Navistar International	1998	$2.9 million	No specific individual requirement[a]	40,000 heavy-duty diesel engines produced from 1994 to 1998	http://www2.epa.gov/enforcement/navistar-international-transportation-corporation-diesel-engines-settlement
Volvo Truck Corp.	1998	$5 million	$9 million for environmental projects	10,000 heavy duty diesel engines produced from 1994 to 1998	http://www2.epa.gov/enforcement/volvo-truck-corporation-diesel-engines-settlement
Ford Motor Company	1998	$2.5 million	$1.3 million to modify affected	60,000 MY1997 vans	http://www2.epa.gov/sites/production/files/20

Company	Year	Civil Penalties	Additional Relief	Affected Vehicles/Equipment	Link
			vehicles; plus $4.0 million to purchase nitrogen oxide credits and to support environmental projects		14 06/documents/defeat.pdf http://www2.epa.gov/enforcement/ford-motor-company-clean-air-act-settlement
American Honda Motor Co.	1998	$12.6 million	$250 million to extend emissions warranties and provide maintenance; plus $4.5 million in other environmental projects	1.6 million MY1995-MY1997 passenger vehicles	http://www2.epa.gov/enforcement/american-honda-motor-company-clean-air-act-settlement
General Motors Corp.	1995	$11 million	$25 million recall and retrofit; plus $9 million in other actions (may include older vehicle buyback or purchasing new school buses)	500,000 MY1991-MY1995 passenger vehicles	http://echo.epa.gov/enforcement-case-report?id=HQ-1996-0001 http://www.justice.gov/archive/opa/pr/Pre_96/November95/596.txt.html
Volkswagen	1974	$120,000	None	25,000 MY1973 passenger vehicles	http://www.autosafety.org/sites/default/files/imce_staff_uploads/VW%20Defeat%20Device%20$120,00%20fine%203-12-74%20Pr.pdf
Chrysler, Ford, General Motors, and Toyota	1973	None	Remove ambient temperature sensors from new vehicles		

Source: Congressional Research Service, with data from the U.S. Environmental Protection Agency and various news sources (see hyperlinks included in the table).

Notes: MY refers to "model year." Additional remediation/injunctive relief may include actions not directly involving the affected vehicles. For example, other potential NO_x emissions reductions (e.g., providing low-emission buses for municipal fleets could qualify).

[a] Plus a share of $850 million combined investment from heavy-duty engine manufacturers involved in the settlement. These investments were to cut emissions from new engines 15 months in advance of new emissions standards.

CONGRESSIONAL RESPONSE

What Is the Role of Congress?

Congress' initial role was to establish the anti-defeat device provisions in the 1970 amendments to the CAA. On October 8, 2015, representatives of VW and EPA testified on

health-based air quality standards, address interstate and international emissions transport, and restore visibility in national parks and wilderness areas.

Notwithstanding the undisputed air quality progress since 1970, several challenges remain to reduce pollution in areas exceeding federal standards and to continue protection in areas meeting the standards. For example, the movement of air pollutants across state borders poses difficulties for states whose air quality is affected by emissions from upwind jurisdictions. Some states are concerned about emissions transported from other countries.

MEMORANDUM'S DIRECTIVES

The memorandum presents air quality management goals—such as timely petition reviews—but does not prescribe particular methods. This section summarizes the directives in the context of the statutory provisions.

AIR QUALITY STANDARDS

The Clean Air Act establishes roles for federal and state agencies. EPA establishes minimum national air quality standards—National Ambient Air Quality Standards (NAAQS)—and is required to review them every five years. States adopt State Implementation Plans (SIPs) to ensure compliance with the standards. EPA reviews SIPs to ensure they meet statutory requirements. The act requires EPA, under certain conditions, to impose sanctions and to issue a Federal Implementation Plan (FIP) if a state fails to submit or implement an adequate SIP.

The White House memorandum directs EPA to review the multi-step process for setting and revising NAAQS—in particular, to assess the transparency of the science reviews, develop guidance differentiating science and policy considerations, and seek Clean Air Scientific Advisory Committee advice regarding background pollution and potential adverse effects from NAAQS compliance strategies. The memorandum is silent on prior Administrations' reviews, which have considered the "timeliness, scientific integrity and transparency" of the NAAQS review process.

NAAQS IMPLEMENTATION AND PERMITTING PROGRAMS

The memorandum directs EPA to review related "existing rules, guidance, memoranda, and other public documents," including preconstruction permit applications. It directs EPA to determine whether to revise or rescind any of these documents, consistent with the law, and to improve the timeliness of SIP and permit decisions. Some of the memorandum's timeline goals reiterate EPA's statutory deadlines (e.g., for SIP approvals and permit reviews). The memorandum's timeline goal for preconstruction permit issuance may be less relevant, however, given that states issue most of these permits.

The memorandum directs EPA to issue NAAQS implementation regulations and guidance concurrent with the NAAQS revisions. Historically, EPA has issued implementation regulations and guidance, in collaboration with states, subsequent to the NAAQS revisions.

International Emissions and Exceptional Events

The act authorizes EPA to provide regulatory relief, under certain conditions, for air emissions deemed beyond the control of a state. For example, monitoring data from exceptional events—unusual or naturally occurring events affecting air quality—may be excluded from determinations of NAAQS compliance status ("Section 319(b) petitions"). EPA may also approve SIPs that demonstrate that an area would have met NAAQS if not for emissions from another country ("Section 179B petitions"). The memorandum directs EPA to ensure timely processing of these petitions; the statute does not specify deadlines for them.

The memorandum directs EPA to consider Section 179B petitions from any state—not just those bordering Canada and Mexico—and to consider whether emissions from Asia affect domestic air quality. EPA has previously considered international transport: The 2015 ozone rule discussed ozone precursor emissions that can contribute to "background" levels and described outreach to other countries, including China (p. 65443). While Section 179B does not limit the petitions to particular states, EPA's 2016 comment solicitation suggested that international ozone may have greatest influence in areas "in the immediate vicinity of Mexico or Canada."

The memorandum's directives also address use of modeling data and, in the context of intrastate and regional emissions trading, development and implementation of offset policies in rural areas.

Regional Haze

The White House memorandum directs EPA to review FIPs issued under the Regional Haze program, which limits pollution impairing visibility in certain national parks and wilderness areas, and to replace such FIPs with approvable SIPs. EPA previously announced plans to review a 2017 Regional Haze rule but has not specified whether it will address the memorandum's directives.

ISSUES FOR CONGRESSIONAL CONSIDERATION

The effect of the President's memorandum is unclear, given that it does not specify changes EPA is to make. Separately, EPA has announced initiatives related to the memorandum's goals: EPA established an Ozone Cooperative Compliance

Task Force, issued new guidance for preconstruction permitting, proposed new requirements for use of scientific studies in rulemaking, and plans to issue new exceptional events guidance.

Issues for Congress may include

- The effect of EPA's initiatives on federal-state partnerships, including how they would affect EPA's statutory obligation to set health-based air quality standards and oversee state implementation.
- Whether the act provides sufficient clarity about international pollution impacts on domestic air quality or if this determination should remain at EPA's discretion. It is unclear how EPA will define *background emissions* in future actions while it pursues its stated goal to maximize flexibility of Section 179B petitions and other regulatory relief tools.
- Proposals to modify the statute based on reviews of the NAAQS process and other programs.

PART 2. THE CLEAN POWER PLAN

In: The Clean Air Act and The Clean Power Plan
Editor: Claire Ferguson
ISBN: 978-1-53614-825-1

Chapter 8

RECONSIDERING THE CLEAN POWER PLAN*

James E. McCarthy

SUMMARY

On October 10, 2017, the U.S. Environmental Protection Agency (EPA) proposed to repeal the Clean Power Plan (CPP), an Obama Administration rule that would limit carbon dioxide (CO_2) emissions from existing fossil-fuel-fired power plants. Because power plant CO_2 emissions account for about 30% of total U.S. anthropogenic emissions of greenhouse gases (GHGs), the CPP has been seen as the most important U.S. regulation addressing climate change.

The CPP has not gone into effect: In February 2016, the U.S. Supreme Court stayed its implementation pending the completion of judicial review. Even had it not been stayed, the rule's limits on CO_2 emissions were not scheduled to begin taking effect until 2022. The Court's action delayed various planning requirements that would have determined how states intended to structure compliance with the rule's overall objectives.

Unlike the suspension of the CPP that is currently in place due to the Supreme Court's stay, repealing a promulgated rule requires that the promulgating agency go through the same steps as the original rulemaking, a process governed in this case by Section 307(d) of the Clean Air Act. The first step in the repeal process is a 60-day comment period following publication of the proposed repeal in the *Federal Register*. Ultimately, EPA will need to address all significant comments and criticisms that it receives during the public comment period when it promulgates a final decision on the proposed repeal. The agency's decision could then be subject to judicial review.

Although the agency is proposing to repeal the CPP, it did not propose repeal of the GHG "endangerment finding," the 2009 agency finding that emissions of CO_2 and other GHGs endanger public health and welfare. Without addressing the finding, the agency appears to have a continuing obligation to limit emissions of CO_2 from power plants. Thus, in addition to the proposed repeal of the CPP, EPA has prepared and sent for interagency

* This is an edited, reformatted and augmented version of a Congressional Research Service report, Publication No. R44992, dated October 25, 2017.

review an Advance Notice of Proposed Rulemaking (ANPRM) to solicit information on systems of emission reduction that it might require in a future rule to replace the CPP.

The net effect of EPA's repeal and the ANPRM may be a continuing period of regulatory uncertainty for the states and industry. In the meantime, the electric power industry is changing rapidly as a result of several factors, including market forces, state and federal regulations, technological innovation, and federal tax incentives. Many coal-fired power plants are being retired, and the new electric generation replacing those plants is overwhelmingly powered by natural gas or renewable power. Because coal-fired plants emit far more CO_2 per unit of power than their replacements, total emissions of CO_2 from electric power generation declined almost 25% between 2005 and 2016, while gross domestic product grew and the amount of power generated remained essentially unchanged. This observed decline in annual CO_2 emissions from the electric power sector is 77% of the reductions that EPA projected would occur as a result of the CPP.

Members of Congress may have an interest—for legislative and oversight purposes, as potential commenters, and in responding to constituents—in understanding what it is that EPA has proposed to do with regard to the CPP. This chapter provides background on the CPP and its proposed repeal, describes the administrative steps that are required to repeal or amend a rule, and discusses how the CPP and its proposed repeal fit into the context of recent and projected power sector evolution.

INTRODUCTION

On October 10, 2017, the U.S. Environmental Protection Agency (EPA) proposed to repeal the Clean Power Plan (CPP), an Obama Administration rule that would limit carbon dioxide (CO2) emissions from existing fossil-fuel-fired power plants.[1] Because power plant CO_2 emissions account for about 30% of total U.S. anthropogenic emissions of greenhouse gases (GHGs), the CPP has been seen as the most important U.S. regulation addressing climate change.

Congress has taken a keen interest in the CPP. Following its promulgation, Congress passed a resolution of disapproval under the Congressional Review Act in December 2015 that would have overturned the rule. President Obama vetoed the resolution.

The courts and the states have also played key roles: After more than 100 parties, including 46 states, either petitioned the Court of Appeals for the DC Circuit to review the rule or intervened in support of it, the Supreme Court granted applications to stay the rule for the duration of litigation.

Now, EPA has regained the spotlight. With new leadership in the Trump Administration, EPA has proposed to repeal the rule that it had drafted and defended under President Obama.

---

[1] EPA, "Repeal of Carbon Pollution Emission Guidelines for Existing Stationary Sources: Electric Utility Generating Units," Proposed Rule, 82 *Federal Register* 48035, October 16, 2017 (hereinafter, "Proposed Repeal"). Links to an EPA fact sheet and the Regulatory Impact Analysis are at https://www.epa.gov/stationary-sources-air-pollution/ electric-utility-generating-units-repealing-clean-power-plan-0.

As first steps in the repeal process, a 60-day comment period began following publication of the proposed repeal in the *Federal Register*. Following the comment period, EPA must review and respond to "each of the significant comments, criticisms, and new data submitted in written or oral presentations during the comment period."[2] The proposed repeal will likely unleash a torrent of comments: Proposal of the rule itself in 2014 led to 4.3 million public comments—more than EPA had received on any proposed rule in its 45-year history.

Members of Congress may have an interest, as potential commenters and in responding to constituents, in understanding what it is that EPA has proposed. This chapter provides background on the CPP and its proposed repeal, describes the administrative steps that are required to repeal or amend a rule, and discusses how the CPP and its proposed repeal fit into the context of recent and projected power sector evolution.

BACKGROUND

EPA promulgated the CPP on August 3, 2015, under Section 111(d) of the Clean Air Act (CAA).[3] The rule would have required that states limit emissions of CO_2 from fossil-fueled power plants.[4] The rule would have phased in state-specific limits on power plant emissions or emission rates. The targets varied depending on the mix of power sources in each state, but overall EPA projected a 32% reduction in total power sector CO_2 emissions nationwide by the time the rule was fully implemented in 2030.

Interest in the rule reflects what is generally conceded to be the importance of its potential effects. The economy and the health, safety, and well-being of the nation depend on a reliable and affordable power supply, which many have contended would be adversely affected by controls on CO_2 emissions from power plants. At the same time, an overwhelming scientific consensus has formed around the risks, potentially catastrophic, of GHG-induced climate change. Proponents of the rule maintain that the CPP addressed these risks in cost-effective ways that take account of ongoing changes in the electric power sector and that gave states and affected industry significant flexibility in complying with the rule's mandates. Others observe that CO_2 emissions have already declined significantly without the CPP as a result of other factors, including market forces, state and federal

---

[2] 42 U.S.C. 7607 (d)(6)(B).

[3] EPA, "Carbon Pollution Emission Guidelines for Existing Stationary Sources: Electric Utility Generating Units," Final Rule, 80 *Federal Register* 64661, October 23, 2015. Information regarding the rule, including EPA's Regulatory Impact Analysis and numerous EPA fact sheets can be found at https://web.archive.org/web/20161104002205/http://www2.epa.gov/cleanpowerplan/clean-power-plan-existing-power-plants.

[4] See CRS Report R44341, *EPA's Clean Power Plan for Existing Power Plants: Frequently Asked Questions*, by James E. McCarthy et al.

lower carbon sources might be wind or solar power units located miles away from the coal-fired unit whose emissions are to be reduced. The proposed repeal states that such outside-the-fence-line measures are not authorized by Section 111; it maintains that the agency's historical practice has been to interpret the authority in Section 111 to allow only measures that can be applied "at and to an individual source" of pollution.[8]

In addition to the proposed repeal of the CPP, EPA has prepared and sent for interagency review an Advance Notice of Proposed Rulemaking (ANPRM) "to solicit information on systems of emission reduction that are in accord with the legal interpretation proposed in this notice (i.e., those that are applicable at and to an individual source)."[9] This would appear to be an acknowledgement that the agency may need to do more than repeal the CPP without committing to any specific regulatory approach. EPA issued a finding in 2009 that GHG emissions endanger public health and welfare, a finding that it has not proposed to repeal. Because fossil-fueled electric power plants are the largest single source of U.S. anthropogenic GHG emissions, the CAA could arguably be said to require the agency to set standards for those emissions.

In setting emission standards, Section 111 requires standards of performance that reflect "the degree of emission limitation achievable through the application of the best system of emission reduction." While the term *system* is not further defined, a standard that meets EPA's current interpretation (i.e., that uses only measures applied at and to individual sources) might include requiring improved heat rates (i.e., greater efficiency) at individual fossil-fueled power plants or the use of cleaner fuels, such as natural gas in place of coal. In the past decade, more than 50 coal-fired power plants have switched to natural gas. Although there may be some economic and technical constraints on the choice of this option, natural-gas-fired power plants emit as much as 50% less CO_2 per unit of power produced.[10]

Going Forward

The net effect of EPA's repeal and the ANPRM may be a continuing period of regulatory uncertainty for the states and industry. In the meantime, due to market forces, state and federal regulations, technological innovation, and federal tax incentives, the electric power industry is changing rapidly.

Market forces have included:

---

[8] For additional discussion of EPA's legal reasoning, see CRS Legal Sidebar LSB10016, *EPA Proposes to Repeal the Clean Power Plan*, by Linda Tsang.

[9] Proposed Repeal, 82 *Federal Register* 48036.

[10] See CRS Report R44090, Life-Cycle Greenhouse Gas Assessment of Coal and Natural Gas in the Power Sector, by Richard K. Lattanzio.

- the abundance and low price of natural gas,
- the flattening of demand for electric power, and
- advances that have sharply lowered the costs of renewable power.

These forces have resulted in the retirement of dozens of coal-fired power plants, their replacement by natural-gas-fired and renewable generation, and a decline in GHG emissions from the power sector.

The market forces have been buttressed by state and federal regulations—principally, the renewable power requirements in place in about 30 states; the caps on GHG emissions in California and nine Northeastern states;[11] efficiency standards set in both state and federal regulations; emission standards limiting cross-state air pollution, mercury, and air toxic emissions; and standards for the disposal of coal combustion waste.

At the same time, significant technological innovations have been deployed, including high-efficiency gas turbines, which result in less CO_2 per unit of power produced for new plants.

In addition, over the last decade, Congress has provided federal tax incentives for the use of wind and solar generation technologies.

As a result of this combination of factors, between 2005 and 2016, emissions of CO_2 from electric power generation declined almost 25%, while gross domestic product grew and the amount of power generated remained essentially unchanged. The CO_2 emission reduction already achieved represents 77% of the reduction that EPA expected the electric power sector to achieve by 2030 under the CPP (See *Figure 1*).

Projections of future emissions from the electric power sector vary depending on numerous assumptions: The Department of Energy's Energy Information Administration (EIA), for example, has projected electricity sector GHG emissions to be 36% below 2005 levels in 2030 with the CPP and other assumptions in place; without the CPP, EIA projects a 22% decline from 2005 levels. A 22% reduction from 2005 levels would be an *increase* from the current level of emissions.[12] Others see a continuation of market forces and state regulation contributing to a decline in GHG emissions even without the CPP, although most likely not as great a decline as what would be achieved under the rule.[13]

Whatever the projection, it is clear that the changes to the industry over the last decade, while substantial, have been uneven. Electricity generation is influenced by state regulation, which in turn is influenced by state resources and the limitations of the transmission grid. California, Texas, Iowa, and a number of other states have seen rapid growth of renewables (especially wind), but other states remain committed to coal.

[11] See CRS Report R41836, The Regional Greenhouse Gas Initiative: Lessons Learned and Issues for Congress, by Jonathan L. Ramseur.

[12] EIA, Annual Energy Outlook 2017.

[13] See, for example, Rhodium Group, "What the CPP Would Have Done," October 9, 2017, http://rhg.com/notes/whatthe-cpp-would-have-done.

Natural gas use for power generation has grown rapidly, but it has grown more in states with natural gas resources and more developed pipeline infrastructure.

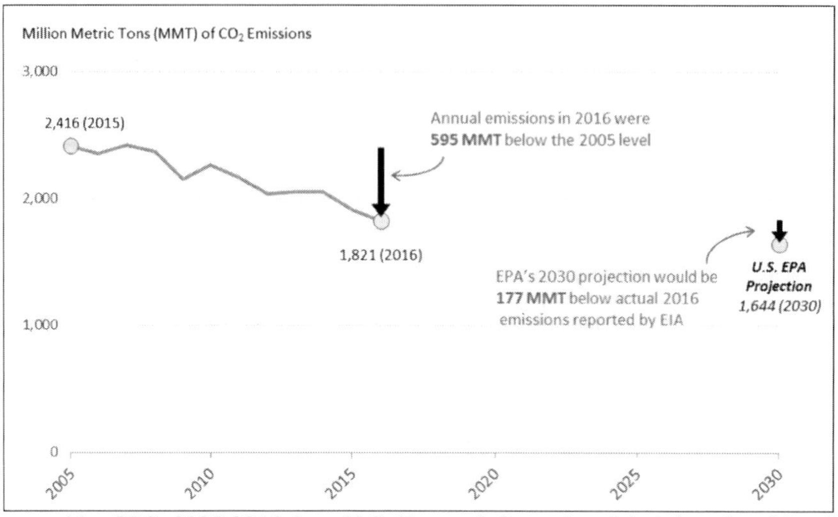

Source: Prepared by CRS; 2005-2016 data: U.S. Energy Information Administration, Monthly Energy Review, September 2017; 2030 projection: EPA, "Carbon Pollution Emission Guidelines for Existing Stationary Sources: Electric Utility Generating Units," Final Rule, 80 Federal Register 64661, October 23, 2015, EPA data converted to metric tons by CRS.

Figure 1. GHG Emissions from Electric Power Generation, 2005-2016.

The CPP was an effort to continue the trend toward lower CO_2 emissions evident in the last decade and to spread the lower emissions associated with renewables, natural gas, and efficiency more evenly among all the states. It set individual state emission targets. The targets would allow higher CO_2 emissions in states currently dependent on coal-fired power but require those states to make greater percentage reductions in emissions.

Without the CPP, the power industry would likely still undergo change as the result of market forces and the need to update or replace older generation facilities. But there would be greater differences among the states and, most likely, less overall reduction in GHG emissions than there would be in a CPP world.

In: The Clean Air Act and The Clean Power Plan
Editor: Claire Ferguson
ISBN: 978-1-53614-825-1

Chapter 9

CLEAN POWER PLAN: LEGAL BACKGROUND AND PENDING LITIGATION IN *WEST VIRGINIA V. EPA**

Linda Tsang and Alexandra M. Wyatt

SUMMARY

On October 23, 2015, the U.S. Environmental Protection Agency (EPA) published its final Clean Power Plan rule (Rule) to regulate emissions of greenhouse gases (GHGs), specifically carbon dioxide (CO_2), from existing fossil fuel-fired power plants. The aim of the Rule, according to EPA, is to help protect human health and the environment from the impacts of climate change. The Clean Power Plan would require states to submit plans to achieve state-specific CO_2 goals reflecting emission performance rates or emission levels for predominantly coal- and gas-fired power plants, with a series of interim goals culminating in final goals by 2030.

The Clean Power Plan has been one of the more singularly controversial environmental regulations ever promulgated by EPA, and the controversy is reflected in the enormous multi-party litigation over the Rule ongoing in the U.S. Court of Appeals for the District of Columbia Circuit (D.C. Circuit). Numerous petitions challenging the Clean Power Plan have been consolidated into one case, *West Virginia v. EPA*. While the litigation is still ongoing at the circuit court level, an unusual mid-litigation application to the Supreme Court resulted in a stay of the Rule, meaning that the Rule does not have legal effect at least for the duration of the litigation. On September 27, 2016, the en banc (full court) D.C. Circuit heard oral argument for the case.

This chapter provides legal background on the Rule, its Clean Air Act (CAA) framework under Section 111, and climate-related lawsuits that have preceded the present litigation over the Clean Power Plan. It then gives an overview of the participants in the current litigation, including two groups of Members of Congress, who have offered briefs in support of both sides. This chapter highlights the major events in the litigation as of the

* This is an edited, reformatted and augmented version of a Congressional Research Service report, Publication No. R44480, dated March 8, 2017.

date of publication, including the Supreme Court stay and oral argument, and the likely timetable of events in the near term.

Some of the main arguments on the merits are then briefly summarized and excerpted from court filings, including

- the standard of review to apply to EPA's action;
- the scope of EPA's overall authority under CAA Section 111;
- whether Section 111 allows the Clean Power Plan's inclusion of generation-shifting, such as from coal-fired power plants to lower-emitting sources of electricity;
- the interpretation of a statutory exclusion in CAA Section 111 that cross-references CAA Section 112's regulation of hazardous air pollutants, particularly in light of the apparent enactment in 1990 of differing House and Senate amendments to the same cross-reference;
- constitutional arguments relating to federalism and separation of powers;
- record-based challenges to the achievability and reasonableness of the Rule; and
- arguments regarding rulemaking procedures.

This chapter concludes with a brief look at parallel litigation in the D.C. Circuit, consolidated as *North Dakota v. EPA*, which is challenging a related EPA regulation that imposes new source performance standards (NSPSs) limiting CO_2 emissions from new, modified, and reconstructed fossil fuel-fired power plants.

INTRODUCTION

On October 23, 2015, the U.S. Environmental Protection Agency (EPA) published its final Clean Power Plan rule (CPP or Rule) to regulate emissions of greenhouse gases (GHGs), specifically carbon dioxide (CO_2), from existing fossil fuel-fired power plants.[1] The goal of the Rule, according to EPA, is to help protect human health and the environment from the impacts of climate change.[2] The CPP has been one of the more singularly controversial environmental regulations ever promulgated, and the controversy is reflected in the enormous multi-party litigation over the Rule in the U.S. Court of Appeals for the District of Columbia Circuit (D.C. Circuit). Numerous petitions challenging the Rule have been consolidated under the caption *West Virginia v. EPA*.[3] While the litigation is still ongoing at the circuit court level, an interlocutory—that is, mid-

---

[1] For more information on the CPP, see CRS Report R44341, EPA's Clean Power Plan for Existing Power Plants: Frequently Asked Questions, by James E. McCarthy et al., and CRS Report R44145, EPA's Clean Power Plan: Highlights of the Final Rule, by Jonathan L. Ramseur and James E. McCarthy.

[2] Carbon Pollution Emission Guidelines for Existing Stationary Sources: Electric Utility Generating Units, Final Rule, 80 Fed. Reg. 64,661, 64,663 (October 23, 2015) (hereinafter "Clean Power Plan Final Rule").

[3] See docket for *West Virginia, et al v. EPA*, No. 15-1363 (D.C. Cir. docketed October 23, 2015). The docket is available electronically through Public Access to Court Electronic Records (PACER) site access; in addition, several websites have been maintaining compilations of major filings and orders in the case, including the Chamber of Commerce of the United States, a petitioner, at its U.S. Chamber Litigation Center, http://www.chamberl itigation.com/ chamber-commerce-et-al-v-epa-esps-rulee; and the Environmental Defense Fund, an intervenor in support of EPA, at its Clean Power Plan Case Resources website at https://www.edf.org/climate/clean-power-plan-case-resources.

litigation—application to the Supreme Court resulted in a stay, or pause, of the Rule.[4] This chapter provides legal background on the Rule, its Clean Air Act (CAA) framework under Section 111, and some of the main climate-related lawsuits that have preceded the present litigation over the CPP. It then gives an overview of the participants in the litigation, including Members of Congress, who have offered briefs in support of both sides. This chapter explains the major events in the litigation as of the date of publication, including the Supreme Court stay and the D.C. Circuit oral argument, and the expected schedule of events in the near term. It then presents condensed summaries of some of the main arguments on the merits. This chapter concludes with a brief look at parallel litigation in the D.C. Circuit that is challenging a related EPA regulation, which limits GHG emissions from new, modified, and reconstructed power plants.

LEGAL BACKGROUND OF THE CLEAN POWER PLAN RULE

Climate Litigation under Other Provisions of the Clean Air Act

The CAA encompasses a number of program authorities, all with the general aim of protecting human health and the environment from emissions that pollute ambient air.[5] Debate over the use of the CAA to regulate GHG emissions has its origins at least as far back as 1999, when several groups filed a petition urging EPA to regulate GHG emissions from new motor vehicles and motor vehicle engines under CAA Section 202.[6] EPA denied the petition in 2003 after soliciting public comments.[7] Shortly thereafter, some of the groups were joined by 12 states and others in filing a petition for review of EPA's decision in the D.C. Circuit.[8] The D.C. Circuit, in a fractured opinion, deferred to EPA's denial of the petition.[9] On appeal, however, in its 2007 decision in *Massachusetts v. EPA*, a five-Justice majority of the Supreme Court held that EPA has statutory authority to regulate GHG emissions under CAA Section 202(a)(1), which requires the EPA Administrator to set emission standards for "any air pollutant" from motor vehicles "which in his judgment

---

[4] See infra, "Supreme Court Order Staying the Clean Power Plan Rule."

[5] 42 U.S.C. §§7401-7628. For an overview of the CAA, *see* CRS Report RL30853, *Clean Air Act: A Summary of the Act and Its Major Requirements*, by James E. McCarthy and Claudia Copeland.

[6] International Center for Technology Assessment et al., Petition for Rulemaking and Collateral Relief Seeking the Regulation of Greenhouse Gas Emissions from New Motor Vehicles Under Article 202 of the Clean Air Act (October 20, 1999), *available at* https://www.regulations.gov/#!documentDetail;D=EPA-HQ-OAR-2001-0002-0001; CAA Section 202, 42 U.S.C. §7421. For more information, *see* CRS Report R40506, *Cars, Trucks, Aircraft, and EPA Climate Regulations*, by James E. McCarthy and Richard K. Lattanzio.

[7] Control of Emissions from New Highway Vehicles and Engines; Notice of Denial of Petition for Rulemaking, 68 Fed. Reg. 52,922, 52,922-33 (September 8, 2003) (citing, *inter alia*, Memorandum from Robert E. Fabricant, EPA Gen. Counsel, to Marianne L. Horinko, EPA Acting Admin., "EPA's Authority to Impose Mandatory Controls to Address Global Climate Change Under the Clean Air Act" (August 28, 2003)).

[8] *Massachusetts v. EPA*, 415 F.3d 50, 53 (D.C. Cir. 2005).

[9] *Id.* at 58-59, 61.

cause[s], or contribute[s] to" air pollution which "may reasonably be anticipated to endanger public health or welfare."[10] GHGs, the Court said, unambiguously fell within the broad definition of "air pollutant."[11] The Court also found that EPA had acted arbitrarily and capriciously in explaining its denial of the petition.[12]

Citing the *Massachusetts v. EPA* decision, EPA issued an "endangerment" finding and a "cause or contribute" finding in December 2009.[13] These findings formed the basis for the light-duty vehicle GHG emission standards and corporate average fuel economy (CAFE) standards issued jointly by EPA and the National Highway Traffic Safety Administration (NHTSA) in 2010.[14]

In 2011, in *American Electric Power Co. v. Connecticut* ("*AEP*"), the Supreme Court unanimously held that EPA's authority to regulate GHG emissions under the CAA—including its power under Section 111(d), the basis of the CPP—displaced any common law tort or nuisance claims against power plants and other GHG emissions sources.[15] The Court in *AEP* explicitly ruled that "air pollutant" includes GHGs when applied to power plants under Section 111, as under Section 202 for motor vehicles.[16] The Court concluded that federal judges may not set limits on GHG emissions because the CAA "empower[s] EPA to set the same limits," and therefore did not allow the plaintiffs, including states, to proceed with their lawsuits against power plant operators.[17]

With GHGs being regulated under CAA Section 202, EPA proceeded with regulating GHGs under other CAA authorities for stationary sources. In particular, EPA interpreted the mobile source GHG regulations as triggering regulations under the Prevention of Significant Deterioration (PSD) program.[18] The PSD program generally requires new or modified stationary sources that will emit threshold amounts (250 or 100 tons per year depending on the type of source) of air pollutants subject to regulation under the CAA to obtain permits and comply with emissions limitations that reflect the "best available control technology" (BACT).[19] EPA likewise sought to regulate GHGs under the Title V permit

---

[10] *Massachusetts v. EPA*, 549 U.S. 497, 531 (2007) (interpreting CAA Section 202(a)(1), 42 U.S.C. §7521).

[11] *Id.* at 528-32.

[12] *Id.* at 532-35.

[13] Endangerment and Cause or Contribute Findings for Greenhouse Gases Under Section 202(a) of the Clean Air Act; Final Rule, 74 Fed. Reg. 66,496 (December 15, 2009).

[14] Light-Duty Vehicle Greenhouse Gas Emission Standards and Corporate Average Fuel Economy Standards; Final Rule, 75 Fed. Reg. 25,324 (May 7, 2010).

[15] *Am. Elec. Power Co., Inc. v. Connecticut*, 564 U.S. 410, 423-29 (2011). "The test for whether congressional legislation excludes the declaration of federal common law is simply whether the statute 'speak[s] directly to [the] question' at issue.... [T]he [CAA] 'speaks directly' to emissions of carbon dioxide from the defendants' plants." *Id.* at 424 (citations omitted).

[16] *Id.* at 424; *see also id.* at 425 ("EPA is currently engaged in a § [1]11 rulemaking to set standards for greenhouse gas emissions from fossil-fuel fired powerplants.... The Act itself thus provides a means to seek limits on emissions of carbon dioxide from domestic powerplants—the same relief the plaintiffs seek by invoking federal common law. We see no room for a parallel track.").

[17] *See id.* at 429.

[18] CAA Sections 160-169, 42 U.S.C. §§7470-7479.

[19] *See generally id.* As summarized by the Supreme Court, "EPA interprets the PSD provisions to apply to sources located in areas that are designated attainment or unclassifiable for any [criteria] pollutant, regardless of whether

program. Title V requires permits for "major sources" with the potential to emit 100 tons per year of any air pollutant.[20] As EPA noted, GHG emissions tend to be "orders of magnitude greater" than emissions of other types of air pollutants, so the statutory thresholds would have swept in many smaller sources not previously subject to CAA permitting.[21] EPA addressed this by issuing a "tailoring rule," structured to phase in GHG permitting under the PSD and Title V programs first for "anyway" sources already subject to permitting for other air pollutants, and then to non-anyway sources meeting higher thresholds.[22]

In 2014, in *Utility Air Regulatory Group v. EPA* ("*UARG*"), the Supreme Court rejected EPA's interpretation of the "triggering" provisions for the stationary source programs; it held that EPA cannot regulate a power plant solely due to its GHG emissions, striking down EPA's "tailoring" rule.[23] Justice Scalia, writing for a five-Justice majority of the Court, stated the following:

> EPA's greenhouse-gas-inclusive interpretation of the PSD and Title V triggers ... [is] unreasonable because it would bring about an enormous and transformative expansion in EPA's regulatory authority without clear congressional authorization. When an agency claims to discover in a long-extant statute an unheralded power to regulate "a significant portion of the American economy," we typically greet its announcement with a measure of skepticism.[24]

On the other hand, in a part of the decision joined by seven Justices, the Supreme Court affirmed EPA's authority under the CAA to require Best Available Control Technology (BACT) to regulate GHG emissions from power plants if the source is regulated for other air pollutants, holding EPA's interpretation of such requirements reasonable.[25] In sum, *UARG* held that EPA "may not treat greenhouse gases as a pollutant for purposes of defining a 'major emitting facility' ... in the PSD context or a 'major source' in the Title V context.... EPA may, however, continue to treat greenhouse gases as a 'pollutant subject to regulation under this chapter' for purposes of requiring BACT for 'anyway' sources."[26] As a practical matter, *UARG* affirmed EPA's ability to regulate roughly 83% of U.S. stationary-source GHG emissions under PSD and Title V permitting programs, and struck

---

the source emits that specific pollutant.... [E]very area of the country has been designated attainment or unclassifiable for at least one [criteria] pollutant; thus, on EPA's view, all stationary sources are potentially subject to PSD review." *Util. Air Regulatory Group [UARG] v. EPA*, 134 S. Ct. 2427, 2436 (2014).

[20] *See* CAA Sections 501-07, 302(j), 42 U.S.C. §§7661-7661f, 7602(j).

[21] *UARG*, 134 S. Ct. at 2436 (quoting Regulating Greenhouse Gas Emissions Under the Clean Air Act: Advance Notice of Proposed Rulemaking, 73 Fed. Reg. 44,354, 44,498-44,499 (July 30, 2008)).

[22] Prevention of Significant Deterioration and Title V Greenhouse Gas Tailoring Rule: Final Rule, 75 Fed. Reg. 31,514, 31,523-25 (June 3, 2010) (e.g., setting, as "second step" of tailoring rule, 100,000 ton per year threshold for new sources, and soliciting comment on "third step").

[23] 134 S. Ct. at 2439-46.

[24] *Id.* at 2444 (quoting *FDA v. Brown & Williamson Tobacco Corp.*, 529 U.S. 120, 159 (2000)).

[25] *Id.* at 2447-49.

[26] *Id.* at 2449.

down its ability to regulate the additional 3% that would have been reached had the tailoring rule been upheld.[27]

Clean Air Act Section 111

In 2011, EPA finalized a settlement agreement with states and others to promulgate New Source Performance Standards (NSPSs) for GHG emissions from fossil fuel-fired power plants under Section 111(b) of the CAA, and emission guidelines covering existing power plants under Section 111(d).[28] President Obama also directed EPA to issue GHG regulations under Section 111(b) and 111(d) in a presidential memorandum issued in June 2013.[29]

As characterized by EPA,[30] Section 111 operates to address one of three "general categories of pollutants emitted from existing stationary sources," the other two being (1) "criteria" air pollutants under the National Ambient Air Quality Standards (NAAQS) program under CAA Sections 108-110;[31] and (2) "hazardous air pollutants" (HAP) under the National Emission Standards for Hazardous Air Pollutants (NESHAP) program under CAA Section 112.[32] Section 111 addresses "air pollution which may reasonably be anticipated to endanger public health or welfare."[33] Section 111 directs EPA to list categories of stationary sources that cause or contribute significantly to such air pollution; to establish NSPSs for new sources within any such category; and then to issue a "procedure" requiring states to submit plans that establish standards of performance for existing sources in a category, under certain conditions. In other words, NSPSs under Section 111(b) may trigger what EPA terms "emission guidelines" under Section 111(d). Portions of CAA Section 111 primarily relevant to the CPP litigation are excerpted below (with indentations and bracketed notations added for readability):

---

[27] See CRS Legal Sidebar WSLG1016, The Supreme Court's Latest Greenhouse Gas Ruling: Good News and Bad News for EPA, by Robert Meltz. Please contact Linda Tsang with any questions regarding this Legal Sidebar.

[28] *See* Settlement Agreement Between State of New York, et al., and U.S. EPA, December 23, 2010, *available at* http://www2.epa.gov/sites/production/files/2013-09/documents/ boilerghgsettlement.pdf; CRS Report R41103, *Federal Agency Actions Following the Supreme Court's Climate Change Decision in Massachusetts v. EPA: A Chronology*, by Robert Meltz, p. 7 (available upon request). Please contact Linda Tsang with any questions regarding this report.

[29] Presidential Memorandum to EPA, "Power Sector Carbon Pollution Standards" (June 25, 2013), *available at* https://www.whitehouse.gov/the-press-office/2013/06/25/ presidential-memorandum-power-sector-carbon-pollutionstandards.

[30] Resp't EPA's Initial Br. at 6, *West Virginia v. EPA*, No. 15-1363 (D.C. Cir. filed March 28, 2016) ("EPA Br."). Hereinafter all citations to briefs are to those filed in the CPP litigation, *West Virginia v. EPA*, and which are available via that case's docket, *supra* footnote 3.

[31] CAA Section 108-110, 42 U.S.C. §§7408-7410.

[32] CAA Section 112, 42 U.S.C. §7412. There can be overlap among the categories in certain ways; Section 111(b) can be used for "criteria" air pollutants, etc.

[33] CAA Section 111(b)(1)(A), 42 U.S.C. §7411(b)(1)(A).

(a) Definitions. For purposes of this section:

(1) The term "standard of performance" means a standard for emissions of air pollutants which reflects the degree of emission limitation achievable through the application of the best system of emission reduction [BSER] which (taking into account the cost of achieving such reduction and any nonair quality health and environmental impact and energy requirements) the Administrator determines has been adequately demonstrated.

[...]

(3) The term "stationary source" means any building, structure, facility, or installation which emits or may emit any air pollutant....

[...]

(b) (1) (A) The Administrator shall ... publish (and from time to time thereafter shall revise) a list of categories of stationary sources. He shall include a category of sources in such list if in his judgment it causes, or contributes significantly to, air pollution which may reasonably be anticipated to endanger public health or welfare.

(B) Within one year after the inclusion of a category of stationary sources in a list under subparagraph (A), the Administrator shall publish proposed regulations, establishing Federal standards of performance [i.e., NSPSs] for new sources within such category....

[...]

(d) (1) The Administrator shall prescribe regulations which shall establish a procedure similar to that provided by section [1]10 of this title [which provides for State Implementation Plans for NAAQS] under which each State shall submit to the Administrator a plan which

(A) establishes standards of performance for any existing source for any air pollutant

I for which air quality criteria have not been issued or which is not included on a list published under section [1]08(a) ...

[From this point, there is dispute in the litigation regarding how subparagraph (i) continues; the House-originated amendment, which appears in both the *U.S. Code* and the *Statutes at Large*, ends subparagraph (i) with "*or emitted from a source category which is regulated under section [1]12*" while the Senate-originated amendment, which appears only in the *Statutes at Large* and not the *U.S. Code*, ends subparagraph (i) with "*or section [1]12(b)*"[34]] but

II to which a standard of performance under this section would apply if such existing source were a new source, and

(B) provides for the implementation and enforcement of such standards of performance.

Regulations of the Administrator under this paragraph shall permit the State in applying a standard of performance to any particular source under a plan submitted under this paragraph to take into consideration, among other factors, the remaining useful life of the existing source to which such standard applies.

(2) The Administrator shall have the same authority—

(A) to prescribe a plan for a State in cases where the State fails to submit a satisfactory plan as he would have under section [1]10(c) of this title in the case of failure to submit an implementation plan, and

(B) to enforce the provisions of such plan in cases where the State fails to enforce them as he would have under sections [1]13 and [1]14 of this title with respect to an implementation plan....[35]

An analysis by the American College of Environmental Lawyers observed that since the 1970s, EPA has promulgated emission guidelines under Section 111(d) of the CAA on seven occasions[36] (as well as six additional occasions in conjunction with the requirements of CAA Section 129,[37] which the 1990 CAA amendments added to specifically require Section 111 NSPS and emission guidelines meeting certain requirements for solid waste incinerators).[38] Air pollutants and source categories for which EPA has issued emission guidelines under Section 111(d) include, among others, methane and non-methane compounds from large landfills; acid mist from sulfuric acid production units; fluoride emissions from phosphate fertilizer plants; reduced sulfur emissions from kraft pulp mills; and fluoride emissions from primary aluminum plants.[39] In addition, EPA's 2005 Clean

[34] For discussion of the discrepant House and Senate amendments to the cross-reference to CAA Section 112, 42 U.S.C. §7412, *see* "Section 112 Exclusion," below.

[35] Excerpted from CAA Section 111, 42 U.S.C. §7411.

[36] American College of Environmental Lawyers (ACOEL), "Memorandum for Environmental Council of the States (ECOS) Concerning Clean Air Act 111(d) Issues," February 22, 2014, 5, 8-10, http://acoel.org/file.axd?file=2014%2f4%2fACOEL+ECOS+FINAL+MEMO+ and+Cover.pdf.

[37] 42 U.S.C. §7429.

[38] *See* ACOEL, footnote 36, at 5-8 (citing 40 C.F.R. Parts Cb, Ce, BBBB, DDDD, FFFF, and MMMM).

[39] *See generally id.* at 8-10 (citing, *inter alia*, 61 Fed. Reg. 9,905 (March 12, 1996); 60 Fed. Reg. 65,387 (December 19, 1995); 45 Fed. Reg. 26,294 (April 17, 1980); 44 Fed. Reg. 29,828 (May 22, 1979); 42 Fed. Reg. 12,022 (March 1, 1977)). Some of these source categories are regulated for other hazardous air pollutants under Section 112. *See, e.g.*, 40 C.F.R. Part 63 (NESHAP), subpts. S (including kraft pulp mills), BB (phosphate fertilizer plants), and AAAA (landfills).

Air Mercury Rule (CAMR) delisted coal-fired power plants from CAA Section 112 and, instead, established a cap-and-trade system for mercury under Section 111(d);[40] the D.C. Circuit vacated CAMR in 2008 on grounds unrelated to its cap-and-trade structure.[41]

EPA finalized Section 111(b) NSPSs for GHG emissions from new, modified, and reconstructed power plants at the same time as the Clean Power Plan.[42] As discussed below, these NSPSs, which must apply for the CPP under Section 111(d) to have effect, are also being challenged in the D.C. Circuit.[43]

Overview of Clean Power Plan

EPA published the proposed "Carbon Pollution Emission Guidelines for Existing Electric Utility Generating Units" on June 18, 2014.[44] The Agency conducted significant outreach to interested parties before the Rule's proposal.[45] EPA continued its outreach after the proposal and held several public hearings[46] and received more than 4.3 million public comments, the most ever for an EPA rule.[47] The CPP, as it became known, was finalized on August 3, 2015, and published in the *Federal Register* on October 23, 2015.[48]

Several Congressional Review Act (CRA) resolutions of disapproval were introduced following receipt of the CPP by Congress, including S.J.Res. 24, which was passed by the Senate on November 17, 2015,[49] and by the House on December 1, 2015.[50] President

---

[40] 70 Fed. Reg. 28,606 (May 18, 2005).

[41] *See generally New Jersey v. EPA*, 517 F.3d 574 (D.C. Cir. 2008). The court found that EPA's delisting of the source category from Section 112 was unlawful and that EPA was obligated to promulgate standards for mercury and other hazardous air pollutants under Section 112. *Id.* at 581-84.

[42] Federal Plan Requirements for Greenhouse Gas Emissions from Electric Utility Generating Units Constructed on or Before January 8, 2014; Model Trading Rules; Amendments to Framework Regulations; Proposed Rule, 80 Fed. Reg. 64,966 (October 23, 2015).

[43] *See infra*, "*North Dakota v. EPA*: Section 111(b) Litigation."

[44] Carbon Pollution Emission Guidelines for Existing Stationary Sources: Electric Utility Generating Units; Proposed Rule, 70 Fed. Reg. 34,830 (June 18, 2014).

[45] Before proposal, according to Bloomberg BNA, "Senior Environmental Protection Agency officials consulted with at least 210 separate groups representing a broad range of interests in the Washington, DC, area and held more than 100 meetings and events with additional organizations across regional offices." *EPA Consulted with Hundreds of Groups on Carbon Rule for Existing Power Plants*, DAILY ENVT. REP., April 8, 2014.

[46] *See id.*

[47] More than 34,000 public submissions on the proposal can be viewed at http://www.regulations.gov/#!docketDetail;D=EPA-HQ-OAR-2013-0602. An interactive map allowing users to search for comments by state officials can be found at http://bipartisanpolicy.org/energy-map/.

[48] Carbon Pollution Emission Guidelines for Existing Stationary Sources: Electric Utility Generating Units, Final Rule, 80 Fed. Reg. 64,661, 64,663 (October 23, 2015).

[49] See 161 Cong. Rec. S7965 (daily ed. November 16, 2015); U.S. Senate, Roll Call Votes, 114th Congress—1st Session, Vote Summary on the Joint Resolution (S.J. Res. 24), http://www.senate.gov/legislative/LIS/roll_call_lists/roll_call_vote_cfm.cfm?congress=114&session=1&vote=00306.

[50] Final Vote Results for Roll Call 650, S.J.Res. 24, http://clerk.house.gov/evs/2015/roll650.xml.

Obama vetoed S.J.Res. 24 on December 18, 2015.[51] Other resolutions and bills have been introduced both for and against EPA regulation of GHGs from power plants.[52]

The CPP is a detailed rule with many definitions and provisions not touched on here, many of which are the subject of specific challenges or defenses in the present litigation. This chapter provides only a basic summary as context for the following discussion of the litigation challenging the Rule.

General Structure

Applying CAA Section 111, EPA determined the "best system of emission reduction" (BSER) for affected electric generating units based on three components, or as EPA calls them, "building blocks":

1) heat rate (i.e., efficiency) improvements at affected power plants,
2) generation shifts among affected power plants (particularly from coal generation to natural gas combined cycle generation), and
3) increased use of renewable energy for electricity generation.[53]

EPA then used the BSER to derive national emission performance rates for each of the two subcategories of power plants affected by the Rule:

1) fossil fuel-fired electric steam generating units, of which coal generation accounts for 94%—oil and natural gas contribute the remainder—and
2) natural gas combined cycle (NGCC) units.[54]

Then, EPA calculated state-specific targets by applying the national rates to each state's baseline generation mix.[55]

States could reach their targets without needing to "comply" with the assumptions in the "building blocks" or the subcategory-specific rates, which are not themselves binding. In general, policies to reach the state-specific targets set by EPA would be determined by state plans.[56] States could use an emission-standards approach or a "state measures" approach and, under the latter, could submit multi-state plans or use a variety of other

---

[51] White House, "Memorandum of Disapproval on S.J.Res. 24," press release, December 18, 2015, https://www.whitehouse.gov/the-press-office/2015/12/19/memorandum-disapproval-sj-res-24. Other CRA resolutions in the 114th Congress to disapprove the CPP included H.J.Res. 67, and H.J.Res. 72, which was ordered to be reported by House Committee on Energy and Commerce on November 18, 2015.

[52] For more information on congressional responses to the CPP, see Alissa Dolan, Congressional Review, in CRS Report R44341, *EPA's Clean Power Plan for Existing Power Plants: Frequently Asked Questions*, by James E. McCarthy et al.

[53] Clean Power Plan Final Rule, *supra* footnote 2, 80 Fed. Reg. at 64,666-67, 64,717-811.

[54] *Id.* at 64,667, 64,811-20.

[55] *Id.* at 64,667, 64,820-26.

[56] *Id.* at 64,667-72, 64,707-09, 64,826-912, 64,941-64 (adding 40 C.F.R. Part 60, Subpart UUUU).

policies or programs.[57] In addition, state plans could measure compliance using an emission rate target, measured in pounds of CO_2 per megawatt-hour (MWh) of electricity, or a mass-based target, measured in tons of CO_2.[58] The CPP also requires state plans to include certain other components and considerations, such as electric reliability.[59]

EPA cannot legally compel a state to submit a Section 111(d) plan. Rather, if a state fails to submit a satisfactory plan by EPA's deadline, CAA Section 111(d) authorizes EPA to prescribe a federal plan for the state.[60] This authority is the same, Section 111(d) says, as EPA's authority to prescribe a federal implementation plan when a state fails to submit a state implementation plan to achieve the NAAQS.[61] EPA published a proposed federal plan for existing power plants, along with models for state plans, at the same time it published the final CPP.[62]

Timeline

The CPP, as promulgated, set a deadline of September 6, 2016, for each state to submit an implementation plan to EPA (or face EPA imposition of a federal plan on sources in the state).[63] In lieu of a completed plan, the CPP authorized a state to make an initial submittal by that date and request up to two additional years to complete its submission.[64] In light of the stay issued in conjunction with the pending litigation challenging the rule, these near-term deadlines lack legal effect. If the Rule is ultimately upheld, then new initial compliance deadlines would have to be set thereafter.[65]

The eight-year interim compliance period for the CPP, as promulgated, begins in 2022 and runs through 2029.[66] The interim period is separated into three steps (2022-2024, 2025-2027, and 2028-2029), each with its own interim goal.[67] Under this timeline, affected power plants would have to meet each of the first, second, and third steps' CO_2 emission performance rates or follow an EPA-approved emissions reduction trajectory designed by the state itself for the eight-year period from 2022 to 2029.[68] The CPP, as promulgated,

---

[57] *Id.*

[58] *Id.*

[59] *Id.*

[60] CAA Section 111(d), 42 U.S.C. §7411(d) (referencing CAA Section 110(c), 42 U.S.C. §7410(c)).

[61] *Id.*; *see also* 80 Fed. Reg. at 64,828, 64,840, 64,855-56, 64,861, 64,881-82.

[62] *See* Federal Plan Requirements for Greenhouse Gas Emissions from Electric Utility Generating Units Constructed on or Before January 8, 2014; Model Trading Rules; Amendments to Framework Regulations; Proposed Rule, 80 Fed. Reg. 64,966 (October 23, 2015). EPA's website provides technical support documents and other explanatory materials on the proposal: http://www2.epa.gov/cleanpowerplan/clean-power-plan-existing-power-plants# federal-plan.

[63] *Id.* at 64,946-47 (adding 40 C.F.R. §§60.5760-65).

[64] *Id.*

[65] *See, e.g.*, Rulemaking to Amend Dates in Federal Implementation Plans Addressing Interstate Transport of Ozone and Fine Particulate Matter, Interim Final Rule, 79 Fed. Reg. 71,663 (December 3, 2014) (after court lifted stay of rule, granting EPA motion to toll deadlines for three years, reflecting length of the litigation).

[66] Clean Power Plan Final Rule, *supra* footnote 2, 80 Fed. Reg. at 64,664-74, 64,708, 65,743-44; *id.* at 64,944-46,64,959-60 (adding 40 C.F.R. §§60.5745, 60.5880).

[67] *See id.*

[68] *See id.*

requires compliance with the state's final goal by 2030.[69] If the Rule is upheld, it is possible that some or all of these later compliance dates could be delayed or adjusted as well.

PRIOR LITIGATION CHALLENGING THE CLEAN POWER PLAN

Challenges to the CPP began well before the final Rule was published in the *Federal Register*. For example, when the Rule was proposed in 2014, Murray Energy Corporation (a coal company) and the states of West Virginia, Alabama, Indiana, Kansas, Kentucky, Louisiana, Nebraska, Ohio, Oklahoma, South Carolina, South Dakota, and Wyoming filed petitions for review in the D.C. Circuit. They made several arguments in the alternative: that the court had authority to, and should, issue an extraordinary writ under the All Writs Act;[70] that EPA's public statements about its legal authority to regulate CO_2 emissions constituted final agency action subject to judicial review; and that the court should strike down the 2011 settlement agreement[71] that EPA reached with several other States and environmental groups setting a timeline for deciding on Section 111 rules for power plants.[72] A panel of the D.C. Circuit rejected these arguments and denied the petitions, saying the following:

> Petitioners are champing at the bit to challenge EPA's anticipated rule restricting carbon dioxide emissions from existing power plants. But EPA has not yet issued a final rule. It has issued only a proposed rule. Petitioners nonetheless ask the Court to jump into the fray now. They want us to do something that they candidly acknowledge we have never done before: review the legality of a proposed rule.... We do not have authority to review proposed agency rules.[73]

Oklahoma also sued to challenge the proposal in federal district court in Oklahoma and did not prevail;[74] the 10th Circuit denied Oklahoma's motion for an injunction pending appeal.[75]

States and energy companies also filed emergency petitions for an extraordinary writ in the D.C. Circuit in August 2015, after EPA had released, but not published, the final CPP.[76] A circuit panel again denied these petitions on September 9, 2015.[77] Petitions for

[69] See id.
[70] 28 U.S.C. §1651.
[71] See supra, footnote 28 and accompanying text.
[72] See generally Order, In re Murray Energy Corp., Nos. 14-1112 et al. (D.C. Cir. June 9, 2015).
[73] *Id.* at 6.
[74] *See* Order and Op., Oklahoma *ex rel. Pruitt v. McCarthy*, No. 4:15-cv-00369-CVE-FHM (N.D. Okla. July 27, 2015).
[75] *See* Order, Oklahoma *ex rel. Pruitt v. McCarthy*, No. 15-5066 (10th Cir. August 24, 2015).
[76] *See* Order, *In re* State of West Virginia, et al., No. 15-1277 (D.C. Cir. September 9, 2015).
[77] *Id.*; *see also* Order (D.C. Cir. August 19, 2015) (denying emergency motion to consolidate with Nos.14-1112, et al.).

panel reconsideration and for rehearing en banc to essentially revive the earlier lawsuits challenging the proposed rule were denied as well, on September 29, 2015.[78]

WEST VIRGINIA V. EPA AND CONSOLIDATED CASES

Participants in the Litigation

The Petitioners and Others Opposing the Clean Power Plan

Parties began filing petitions in the D.C. Circuit challenging the final CPP starting on the day the Rule was published in the *Federal Register*, October 23, 2015.[79] CAA Section 307(b) requires that such petitions for review must be filed in the D.C. Circuit within 60 days after the Rule's publication in the *Federal Register*.[80] The deadline for petitions for review of the CPP was therefore December 22, 2015.

Parties that filed petitions challenging the CPP include 26 states. West Virginia and Texas spearheaded a coalition of 24 state petitioners in filing the lead case. Oklahoma, North Dakota, and Mississippi filed their own petitions.[81] The state of Nevada, while not a petitioner, filed a brief supporting the petitioners, raising the number of states opposing the CPP to 28.[82] On March 7, 2017, the court granted North Carolina's motion to withdraw as a petitioner. See *Figure 1*. Other petitioners challenging the Rule include three labor unions, a number of rural electric cooperatives and an association representing them, more than two dozen industry and trade groups, several nonprofit public policy organizations, and more than two dozen fossil-fuel-related companies and local electric utilities. Other fossil-fuel-related companies have moved to intervene on behalf of the petitioners.[83] In all, more than 100 parties filed dozens of petitions challenging the CPP.[84] All of these petitions

---

[78] *See* Orders, *In re* Murray Energy Corp., Nos. 14-1112 et al. (D.C. Cir. September 29, 2015).

[79] *See* docket for *West Virginia v. EPA*, *supra* footnote 3; Clean Power Plan Final Rule, *supra* footnote 2. As noted above, footnote 3, the briefs in the Clean Power Plan litigation are available online in several locations.

[80] 42 U.S.C. §7607(b).

[81] *See* docket for *West Virginia v. EPA*, *supra* footnote 3. State petitioners are West Virginia, Texas, Alabama, Arizona (Corp. Comm'n), Arkansas, Colorado, Florida, Georgia, Indiana, Kansas, Kentucky, Louisiana (Dep't of Envtl. Quality), Michigan (Atty. Gen.), Mississippi, Missouri, Montana, Nebraska, New Jersey, North Carolina (Dep't of Envtl. Quality), North Dakota, Ohio, Oklahoma, South Carolina, South Dakota, Utah, Wisconsin, and Wyoming.

[82] *See* Br. of *Amici Curiae* State of Nevada and Consumers' Research in Supp. of Pet'rs (filed February 23, 2016) ("Nev. and Consumers' Res. Br. Supp. Pet'rs").

[83] *See* Opening Br. of Intervenors Dixon Bros., Inc., Nelson Bros., Inc., Wesco Int'l, Inc., Norfolk So. Corp., Joy Global Inc., Gulf Coast Lignite Coal., and Peabody Energy Corp. in Supp. of Pet'rs (filed February 23, 2016) ("Br. Intervenors Supp. Pet'rs").

[84] *See* docket for *West Virginia v. EPA*, *supra* footnote 3.

have been consolidated into one case, captioned *West Virginia et al. v. EPA et al.*[85] All petitioners jointly filed two briefs on the merits.[86]

In addition, various amici curiae (non-party "friends of the court") have filed briefs on the merits in support of the petitions challenging the Rule. These include a brief filed by Members of Congress, as discussed below.[87] Also among those who filed briefs as amici curiae are a group of scientists;[88] 166 state and local chambers of commerce and other business associations;[89] several legal foundations;[90] electric utilities;[91] former Public Utility Commissioners;[92] and groups representing women, minorities, seniors, and taxpayers.[93] In total, one intervenor brief and 12 *amicus* briefs were filed in support of the petitioners opposing the CPP.[94]

The Respondents and Others Supporting the Clean Power Plan

Respondents are EPA and its current Administrator, Gina McCarthy, in her official capacity.[95] Parties that have intervened in this case in support of respondents include a coalition of 18 states, the District of Columbia, and five other cities and a county (including some in states that have filed petitions challenging the CPP).[96] Other parties intervening in support of the CPP include regional, state, and municipal utilities and power companies;[97]

---

85 *See id.*

86 Opening Br. of Pet'rs on Core Legal Issues (filed February 19, 2016) ("Pet'rs Br. Core Legal Issues"); Opening Br. of Pet'rs on Procedural and Record-Based Issues (filed February 19, 2016) ("Pet'rs Br. Proc. and R. Issues").

87 *See* Br. for Members of Cong. as *Amici Curiae* in Supp. of Pet'rs (filed February 23, 2016) ("Cong. Br. Supp. Pet'rs"); "Members of Congress," below.

88 *See* Br. for *Amicus Curiae* Scientists in Supp. of Pet'rs (filed April 23, 2016) ("Scientists Br. Supp. Pet'rs").

89 *See* Br. of 166 State and Local Bus. Ass'ns as *Amici Curiae* in Supp. of Pet'rs (filed February23, 2016) ("166 Bus. Ass'ns Br. Supp. Pet'rs").

90 *See* Br. for Se. Legal Found. as *Amicus Curiae* in Supp. of Pet'rs (filed April 22, 2016) ("Se. Legal Found. Br. Supp. Pet'rs"); Br. of *Amici* Pac. Legal Found., Tex. Pub. Policy Found., Morning Star Packing Co., Merit Oil Co., Loggers Ass'n of N. Cal., and Norman Brown in Supp. of Pet'rs (filed February 23, 2016) ("Pac. Legal Found. et al. Br. Supp. Pet'rs"); Br. for *Amicus Curiae* Landmark Legal Found. in Supp. of Pet'rs (filed February 23, 2016) ("Landmark Legal Found. Br. Supp. Pet'rs").

91 *See* Br. of *Amicus Curiae* Mun. Elec. Auth. of Ga. in Supp. of Pet'rs (filed February 23, 2016) ("MEAG Br. Supp. Pet'rs"); Br. for Pedernales Elec. Coop., Inc. as *Amicus Curiae* in Supp. of Pet'rs (filed February 23, 2016) ("Pedernales Br. Supp. Pet'rs").

92 *See* Br. for *Amici Curiae* Former State Pub. Util. Comm'rs in Supp. of Pet'rs (filed February 23, 2016) ("Fmr. PUCs Br. Supp. Pet'rs").

93 *See* Br. of Amici Curiae 60Plus Ass'n, Federalism in Action, Hispanic Leadership Fund, Indep. Women's Forum, Nat'l Taxpayers Union, and Taxpayers Prot. Alliance in Supp. of Pet'rs (filed February 23, 2016) ("60Plus Ass'n et al. Br. Supp. Pet'rs").

94 *See* docket for *West Virginia v. EPA*, *supra* footnote 3.

95 *See* Resp't EPA's Initial Br. (filed March 28, 2016) ("EPA Br."). They are represented by the Department of Justice.

96 *See* Br. for State and Mun. Intervenors in Supp. of Resp'ts (filed March 29, 2016) ("State and Mun. Int. Br. Supp. EPA") (New York, California, Connecticut, Delaware, Hawaii, Illinois, Iowa, Maine, Maryland, Massachusetts, Minnesota, New Hampshire, New Mexico, Oregon, Rhode Island, Vermont, Virginia, and Washington; the District of Columbia; cities of Boulder, Chicago, New York, Philadelphia, and South Miami; and Broward County, Florida).

97 The cities of Austin and Seattle are participating through their municipally owned utilities. *See* Br. of Intervenors Calpine Corp., City of Austin d/b/a Austin Energy, City of Los Angeles (Dep't of Water and Power), City of Seattle (City Light Dep't), Nat'l Grid Generation, LLC, N.Y. Power Auth., Pac. Gas & Elec. Co., Sacramento Muni. Util. Dist., and So. Cal. Edison Co. in Supp. of Resp'ts (filed March 29, 2016) ("Power Cos. Int. Br. Supp. EPA").

more than a dozen nonprofit organizations (including environmental organizations);[98] and several energy industry associations.[99]

Two former EPA Administrators are supporting the CPP as amici curiae: William Ruckelshaus, who headed the agency in 1970, when the CAA was enacted, and again in the 1980s; and William Reilly, the EPA Administrator at the time Congress passed the Clean Air Act Amendments of 1990.[100] Former Secretaries of State and Defense and a Career Diplomat for the State Department also filed a brief supporting the CPP as amici curiae,[101] as did a policy institute;[102] a coalition of medical groups;[103] scientists;[104] grid experts;[105] companies[106] and business[107] and labor groups;[108] faith groups;[109] and a local government coalition comprising the National League of Cities, the U.S. Conference of Mayors, and 54 other cities and localities.[110] As discussed below, Members of Congress also filed a brief in support of the CPP.[111] In total, four intervenor briefs and 18 amici curiae briefs were offered in support of the CPP.[112]

---

[98] *See* Br. of Intervenor Envtl. and Pub. Health Orgs. in Supp. of Resp'ts (filed March 29, 2016) ("Envtl. and Pub. Health Orgs. Int. Br. Supp. EPA") (includes Am. Lung Ass'n, Ctr. for Biological Diversity, Clean Air Council, Clean Wis., Conservation L. Found., Envtl. Def. Fund, NRDC, Ohio Envtl. Council, Sierra Club, and six West Virginia groups).

[99] *See* Br. of Intervenors Advanced Energy Econ., Am. Wind Energy Ass'n, and Solar Energy Industries Ass'n in Supp. of Resp'ts (filed March 29, 2016) ("Advanced Energy Ass'ns Int. Br. Supp. EPA").

[100] *See* Br. of Former EPA Admins. William D. Ruckelshaus and William K. Reilly as *Amici Curiae* in Supp. of Resp'ts (filed March 31, 2016) ("Fmr. EPA Admins. Br. Supp. EPA").

[101] *See* Br. for *Amici Curiae* Madeleine K. Albright, Leon E. Panetta, and William J. Burns in Supp. of Resp'ts (filed April 1, 2016) ("Fmr. Dep't of State and Def. Officials Br. Supp. EPA").

[102] Br. of the Inst. for Policy Integrity at N.Y. Univ. Sch. of Law as *Amicus Curiae* in Supp. of Resp'ts (filed April 1, 2016) ("NYU IPI Br. Supp. EPA").

[103] *See* Br. of the Am. Thoracic Soc'y, Am. Med. Ass'n, Am. Acad. of Pediatrics, Nat'l Med. Ass'n, Am. Coll. of Preventive Med., Am. Coll. of Occupational and Envtl. Med., Nat'l Ass'n for Med. Dir. of Respiratory Care, and Am. Pub. Health Ass'n as *Amici Curiae* in Supp. of Resp'ts (filed April 1, 2016) ("Med. and Pub. Health Grps. Br. Supp. EPA").

[104] *See* Br. of *Amicus Curiae* Climate Scientists in Supp. of Resp'ts (filed April 1, 2016) ("Climate Scientists Br. Supp. EPA").

[105] Br. of *Amici Curiae* Grid Experts Benjamin F. Hobbs, Brendan Kirby, Kenneth J. Lutz, James D. Mccalley, and Brian Parsons in Supp. of Resp'ts (filed April 1, 2016) ("Grid Experts Br. Supp. EPA").

[106] Br. of *Amicus Curiae* Dominion Res., Inc. in Supp. of Resp'ts (filed April 1, 2016) ("Dominion Br. Supp. EPA"); Br. of *Amici Curiae* Amazon.Com, Inc., Apple Inc., Google Inc., and Microsoft Corp. in Supp. of Resp'ts (filed April 1, 2016) ("Tech. Br. Supp. EPA"); Br. of *Amici Curiae* Adobe, Inc., Mars, Inc., IKEA N. Am. Svcs. LLC, and Blue Cross and Blue Shield of Mass., Inc., in Supp. of Resp'ts (filed April 1, 2016) ("Adobe, Mars, Ikea, BCBS Br. Supp. EPA").

[107] *Amici Curiae* Br. of Sustainable Bus. Orgs. in Supp. of Resp'ts (filed April 1, 2016) ("Sustainable Bus. Orgs. Br. Supp. EPA") (includes Am. Sustainable Bus. Council, U.S. Black Chambers, Inc., Integrative Healthcare Policy Consortium, Sustainable Furnishings Council, Nat'l Small Bus. Network, Green Am., Climate Action Bus. Ass'n, and 18 other state, regional, and minority Chambers of Commerce and other business associations and companies).

[108] Br. of *Amicus Curiae* Serv. Emps. Int'l Union in Supp. of Resp'ts (filed April 1, 2016) ("SEIU Br. Supp. EPA").

[109] Br. of *Amici Curiae* Catholic Climate Covenant et al. in Supp. of Resp'ts (filed April 1, 2016) ("Faith Grps. Br. Supp. EPA") (includes 41 Christian and Jewish faith groups).

[110] *See* Br. of *Amici Curiae* Nat'l League of Cities; U.S. Conf. of Mayors; and 54 Cities, Counties, and Mayors in Supp. of EPA (filed April 1, 2016) ("Cities Br. Supp. EPA").

[111] *See* Br. *Amici Curiae* of Current Members of Cong. and Bipartisan Former Members of Cong. in Supp. of Resp'ts (filed March 31, 2016) ("Cong. Br. Supp. EPA"); "Members of Congress," below.

[112] *See also* Corrected Br. of *Amici Curiae* Fmr. State Envtl. and Energy Officials in Supp. of Resp'ts (filed April 1, 2016) ("Fmr. State Envtl. and Energy Officials Br. Supp. EPA"); Br. of Leon G. Billings and Thomas C. Jorling as Amici Curiae in Supp. of EPA (filed April 1, 2016) ("Cong. CAA Staff Br. Supp. EPA"); Br. of *Amici Curiae*

Four states have not joined the litigation: Alaska (for which EPA did not set a goal in the final Rule),[113] Idaho, Pennsylvania, and Tennessee.

Members of Congress

Large groups of Members of Congress have filed amici curiae briefs on both sides of the litigation. A brief opposing the CPP was joined by 34 current Senators and 171 current Representatives in the 114th Congress. The brief argues, among other things, that Congress excluded power plants regulated under CAA Section 112 from "concurrent regulation" under Section 111(d) and that EPA "usurped the role of Congress" through the CPP's "expansive regulatory requirements."[114]

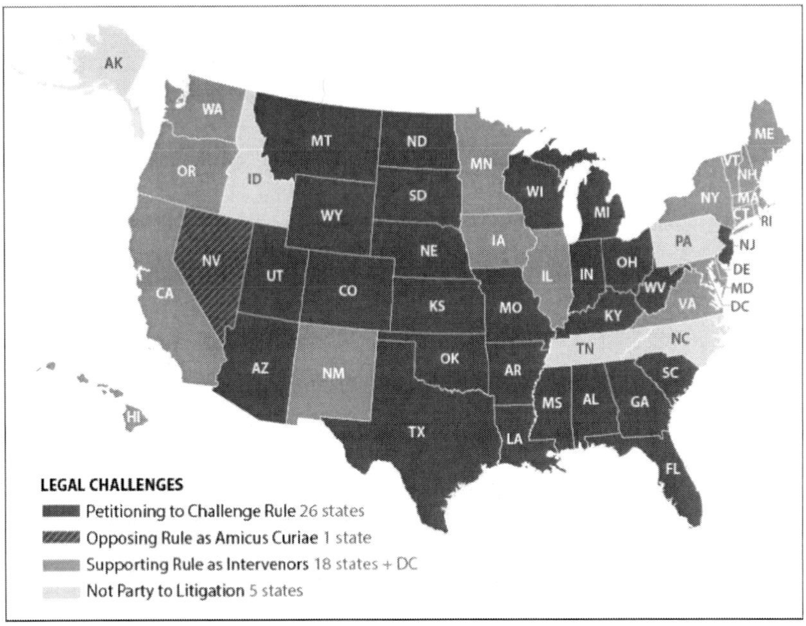

Source: Prepared by CRS from litigation filings in *West Virginia v. EPA*.

Notes: The Clean Power Plan, as finalized, did not set emissions goals for Alaska, Hawaii, Vermont, or the District of Columbia (the latter two because there are no affected electric generating units in those locations).

Figure 1. States Participating in Clean Power Plan Litigation. Consolidated Petitions: *West Virginia et al. v. EPA et al.*, D.C. Circuit No. 15-1363.

Citizens Utility Bd., Consumers Union, and Pub. Citizen, Inc. in Supp. of Resp'ts (filed April 1, 2016) ("Consumer and Ratepayer Orgs. Br. Supp. EPA"); Br. of Union of Concerned Scientists as *Amicus Curiae* in Supp. of EPA (filed April 1, 2016) ("UCS Br. Supp. EPA"); *see generally* docket for *West Virginia v. EPA*, *supra* footnote 3.

[113] *See* Clean Power Plan Final Rule, *supra* footnote 2, 80 Fed. Reg. at 64,664: "Because the EPA does not possess all of the information or analytical tools needed to quantify the BSER for the two non-contiguous states with otherwise affected EGUs (Alaska and Hawaii) and the two U.S. territories with otherwise affected EGUs (Guam and Puerto Rico), these emission guidelines do not apply to those areas, and those areas will not be required to submit state plans on the schedule required by this final action." EPA also did not include Vermont or the District of Columbia in the final Rule because of the lack of affected electric generating units in those locations. *Id.*

[114] *See generally* Cong. Br. Supp. Pet'rs.

A brief in support of the CPP was joined by 44 current and former Senators and 164 current and former Representatives; it argues, among other things, that Congress conferred "broad authority" on EPA in the CAA to help the agency achieve the act's broad anti-pollution objectives, and that the CPP is "consistent with the text, structure, and history" of the CAA.[115]

Major Events in the Litigation

Stay Motions and Scheduling Motions Before the Circuit Court Panel

Many petitioners filed motions to stay the CPP alongside or soon after their petitions for review.[116] Briefing on the stay motions concluded in late December 2015.[117]

On January 21, 2016, the D.C. Circuit panel comprising Judges Srinivasan, Rogers, and Henderson issued an order denying the petitioners' motions to stay the CPP for the duration of the litigation. The court's *per curiam* order denying the motions to stay did not detail the court's reasoning, saying only that "[p]etitioners have not satisfied the stringent requirements for a stay pending court review."[118] However, the case cited in the order, *Winter v. Natural Resources Defense Council*,[119] as well as petitioners' stay motions and respondents' opposition briefs,[120] measured the motions against the four traditional factors for a stay: (1) likelihood of success on the merits, (2) irreparable harm to the movant absent a stay, (3) lack of substantial harm to others if a stay is granted, and (4) public interest. Thus, the stay briefing previewed some of the legal and factual arguments on both sides, including arguments relating to the scope of EPA's authority and the reasonableness of EPA's decisions.

The circuit panel's January 21 order also resolved another matter that had been disputed by the parties: the timing and manner in which the litigation will proceed. The petitioners had jointly requested that the D.C. Circuit bifurcate what they deemed "fundamental issues of legal authority" from record-based challenges to programmatic elements, in order to expedite consideration of the former. The respondents and respondent-intervenors opposed the petitioners' scheduling proposal, urging that the case be briefed and argued in one round addressing all issues. The circuit court's order rejects bifurcation of the case into two phases as requested by petitioners; the order does, however, expedite briefing of the case in order to allow oral argument in the current term of the court.[121] The panel ordered briefing on all issues to be completed in April 2016.[122]

---

[115] *See generally* Cong. Br. Supp. EPA.
[116] *See* docket for *West Virginia v. EPA*, *supra* footnote 3.
[117] *See id.*
[118] Order at 2, *West Virginia v. EPA*, No. 15-1363 (D.C. Cir. January 21, 2016).
[119] 555 U.S. 7, 20 (2008).
[120] *See generally* stay briefs on docket for West *Virginia v. EPA*, *supra* footnote 3.
[121] Order at 2, *West Virginia v. EPA*, No. 15-1363 (D.C. Cir. January 21, 2016).
[122] *Id.*

Supreme Court Order Staying the Clean Power Plan Rule

In a procedurally rare step, various state and industry parties applied to the Supreme Court in late January 2016 for an immediate stay of the Rule, though the circuit court's order was a preliminary decision in a case that is still pending.[123] They submitted their applications to Chief Justice John Roberts, circuit justice for the D.C. Circuit,[124] who referred the actions to the full Court.[125] At the request of the Court, EPA and others provided response briefs in opposition to the applications.[126]

The Court's response was likewise unusual: On February 9, 2016, the Supreme Court issued brief orders granting the applications and staying the Rule.[127] The orders did not provide explanation.[128] The stay pauses the CPP's legal effect while the Rule undergoes judicial review, and EPA may not enforce the Rule for the duration of the stay. (Nevertheless, some states are continuing to plan and prepare, to varying degrees, for the possibility that the Rule will eventually be upheld.)[129]

The Court was split five to four, with Chief Justice Roberts and Justices Scalia, Thomas, Alito, and Kennedy granting the applications, and Justices Ginsburg, Breyer, Sotomayor, and Kagan in favor of denying the applications.[130] The stay order was one of Justice Scalia's last votes on the Supreme Court. Justice Scalia's death on February 13, 2016, and the resulting current vacancy on the Court, will likely affect the course of the CPP litigation—although in ways that are uncertain at present.[131]

---

[123] See CRS Legal Sidebar WSLG1485, Circuit Court Denies Stay of Clean Power Plan; States Ask Supreme Court to Step In (Part 1), by Alexandra M. Wyatt; CRS Legal Sidebar WSLG1489, UPDATED: Circuit Court Denies Stay of Clean Power Plan; States Ask Supreme Court to Step In (Part 2), by Alexandra M. Wyatt.

[124] See Supreme Court of the United States, Circuit Assignments (April 4, 2014), http://www.supremecourt.gov/ about/ circuitAssignments.aspx.

[125] See *West Virginia v. EPA*, Sup. Ct. Docket No. 15A773, available at http://www.suprem ecourt.gov/search. aspx?filename=/docketfiles/15a773.htm.

[126] See id.

[127] 127 See Order in Pending Case, *West Virginia v. EPA*, No. 15A773 (S. Ct. February 9, 2016), https://www.supremecourt.gov/orders/courtorders/020916zr4_4g15.pdf. It does not appear that the Supreme Court has previously stayed or enjoined a final agency rule where a lower court had, after briefing, declined to do so.

[128] See id.

[129] See, e.g., Press Release, State of Connecticut, "Gov. Malloy Statement on Supreme Court's Clean Power Plan Decision" (February 10, 2016), http://portal.ct.gov/Departments_and_ Agencies/Office_of_the_Governor/ Press_Room/Press_Releases/2016/022016/Gov__Malloy_Statement_on_Supreme_Court_s_Clean_Power_Plan _Decision/ ("it is important that we continue with preparations for compliance with the plan while these issues are being worked out in the courts"); Jim Spencer & David Shaffer, Minnesota Vows to Move Ahead with Clean Power, STAR TRIBUNE (February 16, 2016), http://www.startribune.com/minnesota-vows-to-move-ahead-with-clean-power/368563271/. See also E&E Publishing, LLC, POWER PLAN HUB, http://www.eenews.net/ interactive/clean_power_plan (last visited April 13, 2016) (subscription).

[130] *See* Order in Pending Case, *West Virginia v. EPA*, No. 15A773 (S. Ct. February 9, 2016), https://www. supremecourt.gov/orders/courtorders/020916zr4_4g15.pdf.

[131] *See, e.g.*, CRS Legal Sidebar WSLG1495, *What Does Justice Scalia's Death Mean for Congress and the Nation?*, by Kate M. Manuel and Andrew Nolan; Alexandra M. Wyatt, *Environmental Law*, in CRS Report R44419, *Justice Antonin Scalia: His Jurisprudence and His Impact on the Court*, coordinated by Kate M. Manuel, Brandon J. Murrill, and Andrew Nolan.

The D.C. Circuit Court Oral Argument

The three-judge panel originally set to hear *West Virginia v. EPA* comprised Judge Sri Srinivasan, Judge Judith Rogers, and Judge Karen LeCraft Henderson.[132] However, on May 16, 2016, the full D.C. Circuit ordered, "on the court's own motion, that these [consolidated] cases ... be rescheduled for oral argument before the en banc court on Tuesday, September 27, 2016."[133]

The *Federal Rules of Appellate Procedure* provide for hearings en banc in limited circumstances:

> A majority of the circuit judges who are in regular active service and who are not disqualified may order that an appeal or other proceeding be heard or reheard by the court of appeals en banc. An en banc hearing or rehearing is not favored and ordinarily will not be ordered unless:
>
> (1) en banc consideration is necessary to secure or maintain uniformity of the court's decisions; or
> (2) the proceeding involves a question of exceptional importance.[134]

The order was a surprise to many participants and observers.[135] While the order delayed argument by almost four months, it is possible that the ultimate effect will be to expedite resolution of the litigation by essentially skipping the usual step of a panel decision.[136]

The D.C. Circuit en banc court comprises 11 active judges.[137] Two judges did not participate in the decision to hear the case en banc: Chief Judge Merrick Garland and Judge Nina Pillard.[138] Reasons for recusal are generally not given and were not given in this instance. Judge Garland has not been participating in cases and matters while his Supreme Court nomination remains pending. It was unclear, however, why Judge Pillard recused herself from the initial decision to hear the case en banc.

On September 22, 2016 (five days prior to oral argument), the court issued a notice that Judge Pillard will participate in hearing argument.[139] As a result, the en banc court that heard oral arguments in *West Virginia v. EPA* included 10 judges. For oral argument, the

---

132 Order at 2, *West Virginia v. EPA*, No. 15-1363 (D.C. Cir. January 21, 2016).
133 Order at 2, *West Virginia v. EPA*, No. 15-1363 (D.C. Cir. May 16, 2016).
134 Fed. R. App. Proc. Rule 35(a).
135 *See, e.g.*, Ellen M. Gilmer & Robin Bravender, *Clean Power Plan: Latest Legal Twist Shuffles Calendars, Prospects*, E&E NEWS, May 17, 2016, http://www.eenews.net/stories/ 1060037396 ("The court's announcement took everyone by surprise."); Juan Carlos Rodriguez, *3 Takeaways From DC Circ.'s Surprise CPP Argument Pivot*, ENVIRONMENTAL LAW360, May 17, 2016, http://www.law360.com/articles/797357/ 3-takeaways-from-dc-circ-ssurprise-cpp-argument-pivot.
136 *See, e.g., id.*
137 There are a total of 17 judges on the D.C. Circuit—11 active judges and 6 senior judges. *See* U.S. Court of Appeals, the District of Columbia Circuit, *Judges*, https://www.cadc.uscourts. gov/internet/home.nsf/Content/Judges. *See also* D.C. Cir. R. 35.
138 Order at 2, *West Virginia v. EPA*, No. 15-1363 (D.C. Cir. May 16, 2016); U.S. Court of Appeals, the District of Columbia Circuit, *Judges*, https://www.cadc.uscourts.gov/ internet/home.nsf/Content/Judges.
139 Order, *West Virginia v. EPA*, No. 15-1363 (D.C. Cir. September 22, 2016).

> statute that it has long administered and of provisions that go to the core of EPA's mission to protect public health and welfare.
>
> ... Petitioners construe *UARG* as obliterating the second step of *Chevron* in economically and politically significant cases. Under Petitioners' view, ambiguity in such cases must necessarily be resolved against the implementing agency's exercise of its regulatory authority .... But *UARG* does not nullify *Chevron*. *UARG* simply reflected one application of *Chevron* to particular facts, which are readily distinguishable from those here....[153]

A number of EPA's supporting intervenors[154] and amici[155] share the agency's emphasis on deference and "normal administrative law principles."[156]

During the oral argument, the judges questioned whether the court should apply a heightened standard of review beyond the traditional *Chevron* deference,[157] citing *UARG* as an example where the Supreme Court ruled that "[w]e expect Congress to speak clearly if it wishes to assign to an agency decisions of vast 'economic and political significance.'"[158] The judges raised the "major questions" doctrine, citing *King v. Burwell*, in questioning if the CPP was of such "deep 'economic and political significance'" that the court should not presume that Congress delegated interpretive authority and defer to the agency's interpretation of CAA Section 111 unless Congress expressly did so.[159] The court spent a significant portion of the oral argument on whether the CPP was "transformative" in terms of the Rule's impact on the power sector or the method in which EPA has chosen to regulate GHG emissions from power plants.[160] The standard of review applied in this case will likely draw attention to the broader legal debate surrounding the application of *Chevron* deference to an agency's statutory interpretation.[161]

Section 111(d) Scope of Authority

Petitioners focus much of their challenge on EPA's overall design of the CPP and, especially, its inclusion of electricity generation-shifting measures—exemplified by building blocks two and three, discussed above.[162] They allege that this exceeds EPA's

---

153 *Id.* at 40-42.

154 *See, e.g.*, Advanced Energy Ass'ns Int. Br. Supp. EPA at 2-4; Envtl. and Pub. Health Orgs. Int. Br. Supp. EPA at 4.

155 *See, e.g.*, Consumer and Ratepayer Orgs. Br. Supp. EPA at 1.

156 *See* Envtl. and Pub. Health Orgs. Int. Br. Supp. EPA at 3.

157 *See* Ellen M. Gilmer, *supra* footnote 141.

158 *UARG*, 134 S. Ct. at 2444.

159 *See King v. Burwell*, 135 S. Ct. 2480, 2489 (2015). *See also* Philip A. Wallach, *The D.C. Circuit Considers the Clean Power Plan, and Our Constitutional Future*, BROOKINGS INSTITUTE: FIXGOV BLOG (September 27, 2016), https://www.brookings.edu/ blog/fixgov/2016/09/27/the-d-c-circuit-considers-the-clean-power-plan-and-ourconsti tutional-future.

160 *See* James W. Rubin, *Clean Power Plan Finally Has Its Day in Court*, LAW360, October 1, 2016, http://www.law360.com/articles/847138/clean-power-plan-finally-has-its-day-in-court-now-what. *See also* Ellen M. Gilmer, *Clean Power Plan: In Debate over Legal Standard, 'How Transformative is Transformative?'* E&E NEWS, September 29, 2016, http://www.eenews.net/energywire/stories/1060043607.

161 *See e.g.*, CRS Legal Sidebar WSLG1315, *What Does King v. Burwell Mean for Chevron?*, by Daniel T. Shedd.

162 *See generally* Pet'rs Br. Core Legal Issues at 29-61; *supra*, "Overview of Clean Power Plan."

scope of authority under Section 111(d).[163] Arguments that EPA cannot use Section 111(d) to regulate existing power plants at all because power plants regulated for hazardous air pollutants under Section 112 are discussed in the next section.[164]

"Outside the Fenceline" Measures

Petitioners have argued that, for various reasons drawn from both statutory text and context, Section 111 authorizes EPA to require only measures that can be applied to the "performance" of an individual "source" (also known as measures "inside the fenceline"), such as adoption of pollution control devices or other design or operational standards.[165] They say that Section 111 precludes generation-shifting from one type of electric generating unit to another, and does not authorize what they characterize as a reorganization of the nation's electric grid or states' energy economies.[166] Petitioners also maintain that EPA "cannot require States to adopt as a 'standard of performance' reduction obligations that can be met only through *non*-performance by regulated sources,"[167] saying the CPP "does not involve a source improving its emissions performance when it generates, but instead consists of plants *reducing* or *ceasing* work, or nonperformance, as their production is 'shifted' to EPA-preferred facilities."[168] During oral argument, the petitioners argued that power plants would be "subsidizing their competition" because the CPP effectively forces them to buy emission credits from renewable energy providers to meet their emission limits.[169] In their view, EPA has "confuse[d] 'standards of performance' with other [air quality based] programs" in the CAA, such as the NAAQS or acid rain cap-and-trade program.[170] This view is expanded on by various amici curiae,[171] including the Members of Congress supporting the petitioners.[172]

EPA wrote in the preamble to the final CPP that "the phrase 'system of emission reduction' ... is capacious enough to include actions taken by the owner/operator of a stationary source designed to reduce emissions from that affected source, including actions that may occur off-site and actions that a third party takes pursuant to a commercial relationship with the owner/operator."[173] In its brief, EPA states that "[t]he plain meaning of the word 'system' is expansive,"[174] that "statutory context makes clear that the word 'performance' refers to emissions performance, not production performance,"[175] and that

---

[163] *See generally* Pet'rs Br. Core Legal Issues at 29-61.

[164] *See* "Section 112 Exclusion" below.

[165] *See generally* Pet'rs Br. Core Legal Issues at 29-61; *see also, e.g.*, MEAG Br. Supp. Pet'rs at 10.

[166] *See* Pet'rs Br. Core Legal Issues at 29-61.

[167] *See id.* at 24-25.

[168] *See id.*

[169] *See* Philip A. Wallach, *supra* footnote 159.

[170] *See id.* at 54-55 (distinguishing CAA Sections 108-110, 401 et seq., 42 U.S.C. §§7408-7410, 7651 et seq.).

[171] *See, e.g.*, MEAG Br. Supp. Pet'rs at 9-13; Fmr. PUCs Br. Supp. Pet'rs at 4.

[172] *See* Cong. Br. Supp. Pet'rs at 17-19.

[173] Clean Power Plan Final Rule, *supra* footnote 2, 80 Fed. Reg. at 64,761.

[174] EPA Br. at 27; *see also id.* at 46 ("The phrase 'best system of emission reduction' in Section 111(a)(1) contrasts sharply with narrower language appearing elsewhere in the same statutory subsection").

[175] *Id.* at 65.

the agency appropriately applied contextual constraints on BSER by, among other factors, limiting the CPP to actions taken by sources that result in emission reductions from sources.[176] Altogether, EPA says, petitioners "posit limitations on EPA's discretion that are not compelled by the statute, and would frustrate the statutory objective to protect public health and welfare."[177] Several of EPA's supporting intervenors and amici curiae generally agree;[178] for example, intervenor power companies state that "[e]lectricity providers have been shifting generation among affected units and to zero-emitting sources as a means of achieving emission reductions for decades, as these strategies achieve greater reductions at lower cost than by relying on control technology alone."[179]

Part of the oral argument focused on the definition of "system" with respect to BSER. Some of the judges questioned whether EPA's interpretation of BSER went beyond the Section 111(d) boundaries, while other judges noted that EPA's choice of generation-shifting as BSER was a reflection of what has already been demonstrated and implemented for the power sector.[180]

Role of the Federal Energy Regulatory Commission

Petitioners also argue that another limitation on EPA's authority is the authority given by law to the Federal Energy Regulatory Commission (FERC), and that EPA's design of the CPP violates that limitation. They state: "Congress has clearly confirmed the States' plenary authority in this area and granted to a different agency—FERC—the limited federal jurisdiction in this sphere."[181] A brief filed in opposition to the CPP by 18 former state Public Utility Commissioners also contends that the CPP is contrary to the Federal Power Act (FPA),[182] in part because of what they deem EPA's "unprecedented" interpretation of the term "system" in CAA Section 111.[183] They also describe the CPP's effects on states with different regulatory models (vertical integration, restructured, and municipal utilities and electric cooperatives).[184]

EPA counters that the CPP "does not intrude on FERC's power under the Federal Power Act .... The Rule appropriately limits air pollution under the CAA. It does not regulate any kind of electricity sales or rates—interstate or intrastate. Thus, the dividing line between interstate and intrastate rate regulation addressed in the cases cited by

---

[176] *See id.* at 28.

[177] *Id.* at 40.

[178] *See, e.g.*, State and Mun. Int. Br. Supp. EPA at 25-29; Envtl. and Pub. Health Orgs. Int. Br. Supp. EPA at 5-19; Fmr. EPA Admins. Br. Supp. EPA at 16-19; Grid Experts Br. Supp. EPA at 31-37; *see generally* NYU IPI Br. Supp. EPA at 2-31.

[179] *See* Power Cos. Int. Br. Supp. EPA at 2-3. *See also, e.g.*, Dominion Br. Supp. EPA at 10 ("Foreclosing the ability of Dominion and other owners of regulated power plants to rely on trading measures as a means of compliance would unnecessarily increase the Rule's compliance costs....").

[180] *See* Ellen M. Gilmer, *supra* footnote 141.

[181] Pet'rs Br. Core Legal Issues at 38 (citations omitted); *see also* Landmark Legal Found. Br. Supp. Pet'rs at 9-11.

[182] 16 U.S.C. §§791a, et seq.

[183] *See* Fmr. PUCs Br. Supp. Pet'rs at 9-10.

[184] *See id.* at 10-24.

Petitioners has no relevance here."[185] EPA adds that it consulted with FERC and "participated in multiple FERC technical conferences."[186] A group of former state environmental and energy officials, including Public Utility Commissioners, also filed a brief in support of the CPP, arguing in part that "[b]y design, the CPP respects and preserves the fundamental roles of grid operators and the jurisdiction of state regulatory bodies, including environmental agencies and Public Utility Commissions (PUC)."[187]

Section 112 Exclusion

One core set of arguments in the CPP litigation relates to the interpretation of language in CAA Section 111(d)(1)(A), which sets forth exclusions to EPA's authority to issue Section 111(d) emission guideline rules. To understand the dispute, it is useful to briefly review the history of the subsection and its cross-reference to the hazardous air pollutant (HAP) program under CAA Section 112.

Prior to the CAA Amendments of 1990, Section 111(d) required EPA to prescribe regulations for states to submit plans establishing and implementing standards of performance for any existing source, for any air pollutant meeting two requirements: (1) the air pollutant must be one "for which air quality criteria have not been issued or which is not included on a list published under section [1]08(a) or [1]12(b)(1)(A) of this title," and (2) the air pollutant must be one to which a Section 111(b) NSPS would apply if such existing source were a new source.[188] At that time, Section 112(b)(1)(A) described a process for listing HAPs. (Section 108(a) describes a process for listing "criteria" air pollutants; this cross-reference has not changed, nor has the second requirement.) Thus, for any air pollutant to which a Section 111(b) NSPS applied for new sources, EPA had to regulate the same pollutant under Section 111(d) for existing sources *unless* that air pollutant was already listed under the NAAQS or HAP programs.

In 1990, Congress made substantial amendments to CAA Section 112; among other changes, it replaced the former HAP listing process with a list of nearly 200 HAP, now contained in Section 112(b).[189] In doing so, it made Section 111(d)'s cross-reference to Section 112(b)(1)(A) obsolete, as there was no longer an (A). Both the House and the Senate offered amendments to the crossreference—*both* of which were included in the final legislation that was passed, signed into law by President Bush, and included in the *Statutes at Large*. Under the House-originated provision, the Section 111(d) authority applies for any air pollutant that "is not included on a list published under section [1]08(a) of this title or emitted from a source category which is regulated under section [1]12 of this title.... "[190] The House-originated amendment was added to the *U.S. Code* by the House Office of the Law

---

[185] EPA Br. at 59.

[186] *See id.* at 150; *see also id.* at 55, 152.

[187] *See* Fmr. State Envtl. and Energy Officials Br. Supp. EPA at 3-4.

[188] 42 U.S.C. §7411(d)(1)(A) (1988).

[189] 42 U.S.C. §7412(b).

[190] P.L. 101-549, §108(g), 104 Stat. 2399, 2467 (1990), codified at 42 U.S.C. §7411(d)(1)(A)(i).

Revision Counsel.[191] The Senate-originated 1990 amendment to CAA Section 111(d)(1)(A) simply excludes from Section 111(d) regulation any air pollutant "included on a list published under section [1]08(a) or [1]12(b)..."[192] It is not in the *U.S. Code*.

Because power plants are a source category which is regulated under Section 112 for mercury and other HAP,[193] petitioners and their supporters argue that EPA is barred from regulating power plants under Section 111(d) for CO_2 in any manner.[194] Petitioners claim that EPA itself has previously given the *U.S. Code* text its "literal" meaning,[195] and that "EPA's attempts to escape the literal reading of the exclusion are unavailing."[196] In particular, they dispute EPA's "new assertions of ambiguity"[197] and any reliance on the Senate-originated amendment, which they describe as an "erroneous 'conforming amendment.'"[198] Petitioners also cite a footnote in the 2011 Supreme Court case *AEP v. Connecticut*, which said that "EPA may not employ [section 111(d)] if existing stationary sources of the pollutant in question are regulated under ... § [1]12."[199] Members of Congress who submitted an *amicus curiae* brief against the CPP further discuss the Section 112 exclusion and its legislative history,[200] as did petitioners' intervenors, who further argue that EPA's interpretation violates constitutional principles of separation of powers.[201]

EPA generally counters that it "reasonably interpreted" Section 111(d) and its exclusion language, "which is ambiguous in several respects[,] consistent with the Act's purpose, the statutory context, and the legislative history," as well as its own past rulemakings.[202] The Agency states, among other things, that "Petitioners' interpretation of Section 111(d)—which would strip that provision of nearly all effect—is not reasonable, let alone mandatory," and that "when construing [the House-originated amendment] in a particular statutory context, one must take a 'commonsense' approach, and ask not only 'who' is regulated under Section 112 (i.e., source categories including power plants), but also 'what.'"[203] Essentially, EPA interprets Section 111(d)(1)(A) to exclude from Section 111(d)

[191] *See* 2 U.S.C. §§285a-285g (Revision Counsel's authority); *see also* EPA Br. at 89; Pet'rs Br. Core Legal Issues at 69-73.

[192] P.L. 101-549, §302(a), 104 Stat. 2399, 2574 (1990).

[193] EPA has regulated HAP from power plants under CAA Section 112 as part of its mercury and air toxics standards (MATS). The Supreme Court held that EPA's promulgation of the MATS rule was unlawful for failure to properly consider costs at the threshold stage of determining whether such regulation was "appropriate and necessary." *Michigan v. EPA*, 135 S. Ct. 2699, 2707-2711 (2015). It remanded the case to the court of appeals, which remanded the MATS rule without vacatur to EPA to make the additional findings required by the Supreme Court. Order, *White Stallion Energy Ctr. LLC v. EPA*, No. 12-1100 (D.C. Cir. December 15, 2015) (per curiam).

[194] *See, e.g.*, Pet'rs Br. Core Legal Issues at 61-74; Pac. Legal Found. et al. Br. Supp. Pet'rs at 15-20. Note that CO_2 also is not listed as a criteria pollutant under a NAAQS under CAA Section 108(a).

[195] *See* Pet'rs Br. Core Legal Issues at 62-64. *See generally id.* at 61-74 (Section II of brief, "The Section 112 Exclusion Unambiguously Prohibits the Rule").

[196] *See id.* at 64.

[197] *See id.* at 65.

[198] *See id.* at 68-74.

[199] *Id.* at 62 (quoting *AEP*, 131 S. Ct. at 2537 n.7).

[200] *See* Cong. Br. Supp. Pet'rs at 4-13.

[201] *See* Br. Intervenors Supp. Pet'rs at 6-31. *See also, e.g.*, Landmark Legal Found. Br. Supp. Pet'rs at 2-7.

[202] *See generally* EPA Br. at 76-98; *see also* Clean Power Plan Final Rule, *supra* footnote 2, at 64,710-64,715.

[203] *See* EPA Br. 81 (citing *Rush Prudential HMO v. Moran*, 536 U.S. 355, 366 (2002)).

regulation any HAP emitted from a source category regulated under Section 112;[204] otherwise, it says, EPA would have to choose between regulating HAP *or* Section 111 air pollutants, leaving a "gap" and allowing the "unregulated emission of pollutants not listed as 'hazardous' or 'criteria,' but nonetheless dangerous to public health or welfare."[205] In addition, EPA argues that it "properly considered" the Senate-originated amendment as a "clear indication of congressional intent when interpreting Section 111(d),"[206] stating that "[i]t is black-letter law that the U.S. Code cannot prevail over the Statutes at Large when the two are inconsistent."[207] EPA's supporters also weighed in on the Section 112 exclusion issue, including the current and former Members of Congress who filed an amici curiae brief supporting the CPP.[208]

During the oral argument, the judges appeared to grapple with the two differing amendments of Section 111(d) and the complexity and ambiguity of the legislative history.[209] Some judges questioned whether the distinction between "conforming" and "substantive amendments" was confusing the issue further.[210] The judges' questions sought to find an interpretation that would give some effect to both amendments, while ensuring that the resulting interpretation would not create a "loophole" such that sources may evade regulation of certain pollutants.[211]

Constitutional Issues and Canon of Constitutional Avoidance

Petitioners contend that the CPP violates the U.S. Constitution, and that CAA Section 111(d) must be interpreted more narrowly than EPA interprets it so as to avoid certain constitutional issues.[212] For example, petitioners, including the 26 state petitioners opposing the CPP, claim that the CPP impermissibly invades traditional state police powers over the electrical grid and "commandeers" and "coerces" states and their officials and

---

[204] *Id.* (citing Clean Power Plan Final Rule, *supra* footnote 2, 80 Fed. Reg. at 64,713).

[205] *See id.* at 83-84.

[206] *See id.* at 87.

[207] *See id.* at 88 (citations omitted). In general, if there is a discrepancy between the *U.S. Statutes at Large* and the *U.S. Code*, the *U.S. Statutes at Large* is the controlling legal evidence of the law, unless Congress has enacted the relevant title of the *U.S. Code* as positive law; in that case, the *U.S. Code* is also legal evidence of the law. *See* 1 U.S.C. §§12, 204(a). Congress has not enacted the provisions in question as positive law, although there is a bill to do so in the 114th Congress: H.R. 2834 would codify certain laws currently in Title 42 of the *U.S. Code* relating to the environment, including CAA Section 111(d), as a positive law title, which would be a new Title 55 of the *U.S. Code*.

[208] *See* Cong. Br. Supp. EPA at 17-27. *See also, e.g.*, State and Mun. Int. Br. Supp. EPA at 29-33; Envtl. and Pub. Health Orgs. Int. Br. Supp. EPA at 20-27; Cong. CAA Staff Br. Supp. EPA at 24-29.

[209] *See e.g.*, Andrew Childers, *Clean Power Plan Scrutinized Through Lens of Supreme Court*, BLOOMBERG BNA ENERGY & CLIMATE REP., September 27, 2016, http://www.bna.com/clean-power-plan-n57982077646. *See also* Ellen M. Gilmer, *supra* footnote 141; James W. Rubin, *supra* footnote 160.

[210] *See* Juan Carlos Rodriguez, *DC Circ. Puts Clean Power Plan Foes, Backers on Hot Seat*, LAW360, September 27, 2016, http://www.law360.com/articles/845363/dc-circ-puts-clean-power-plan-foes-backers-on-hot-seat.

[211] *See* James W. Rubin, *supra* footnote 160.

[212] *See generally* Pet'rs Br. Core Legal Issues at 78-86.

legislatures.[213] They argue that it does so even with the federal implementation plan option:[214]

> In order to pass constitutional muster, cooperative federalism programs must provide States with a meaningful opportunity to decline implementation. But the Rule does not do so; States that decline to take legislative or regulatory action to ensure increased generation by EPA's preferred power sources face the threat of insufficient electricity to meet demand. The Rule is thus an act of commandeering that leaves States no choice but to alter their laws and programs governing electricity generation and delivery to accord with federal policy.[215]

However, during the oral argument, some judges questioned whether the CPP was any different from other regulatory programs (such as the Americans with Disabilities Act) that impose certain provisions that require state action to implement.[216]

Additionally, in arguing that the court should apply a non-deferential "clear statement" standard of review, petitioners cite D.C. Circuit precedent that "'[f]ederal law may not be interpreted to reach' areas traditionally subject to State regulation 'unless the language of the federal law compels the intrusion' with 'unmistakably clear ... language.'"[217]

Intervenors opposing the CPP expand on several constitutional arguments. In addition to federalism and Tenth Amendment claims,[218] they state that "EPA's attempts to justify the Rule ... trigger a separation-of-powers violation by usurping both the Legislative Branch's lawmaking power and the Judicial Branch's power to 'say what the law is.'"[219] Constitutional arguments are also expanded on by the amici curiae Members of Congress,[220] and by several of the legal groups and other amici opposing the CPP.[221]

EPA, in contrast, defends the CPP as a "textbook example of cooperative federalism."[222] EPA provides, among other reasons, that states can opt to do nothing, in which case the federal plan option imposes no new regulatory obligations on states.[223] The state and municipal intervenors supporting EPA also support the CPP as a lawful implementation of EPA's obligations under the cooperative federalism structure of Section

---

[213] *See generally id.* at 5-6, 78-86.
[214] *See generally id.*
[215] *Id.* at 5-6.
[216] *See* James W. Rubin, *supra* footnote 160; Ellen M. Gilmer, *supra* footnote 141.
[217] *See id.* at 23 (quoting *Am. Bar Ass'n v. FTC*, 430 F.3d 457, 471-72 (D.C.Cir. 2005) (internal quotation marks omitted)).
[218] *See* Br. Intervenors Supp. Pet'rs at 31-38.
[219] *See id.* at 11-12; *see also id.* at 39-43 (arguing that "the purpose of the Constitution's structural divisions of power applies here with special force to prohibit executive overreach and protect individual liberties").
[220] *See* Cong. Br. Supp. Pet'rs at 13-25.
[221] *See, e.g.*, Se. Legal Found. Br. Supp. Pet'rs at 4-10; Pac. Legal Found. et al. Br. Supp. Pet'rs at 24-28; Nev. and Consumers' Res. Br. Supp. Pet'rs at 19-26 (arguing, *inter alia*, that the CPP "undermines democratic accountability").
[222] *See* EPA Br. at 98.
[223] *See generally id.* at 98-106.

111(d), saying, among other things, that "[t]he fact that state regulatory agencies will continue exercising their ordinary oversight over their electric utilities—including over decisions made by power plants to comply with a federal plan—does not mean the Rule commandeers States."[224]

Several amici also dispute constitutional claims against the CPP, including the former EPA Administrators, whose brief describes their view how the CPP fits within the cooperative federalism model.[225] The amici curiae Members of Congress supporting the CPP also counter separation of powers arguments, in part on the grounds that the CAA delegated discretion to EPA with "meaningful criteria" that EPA followed.[226]

Other Arguments Regarding Federal and State Roles

Petitioners and their supporters also base federalism arguments on the text of the CAA.[227] As stated in the brief submitted by 166 business associations, "Section 111(b) grants EPA authority to establish 'standards of performance' for new stationary sources; but Section 111(d) grants the States authority to establish those standards for existing sources. By displacing the authority reserved to the States in setting standards of performance for existing sources ... EPA has violated the statute's unambiguous terms."[228] In addition, as matter of federal law, in the words of the amici curiae brief filed by the Pacific Legal Foundation and others, "[s]ince at least 1964, the national electric power system has been characterized by a 'bright line' divide between federal authority over wholesale sales of power in interstate commerce, regulated by the federal government, and state authority over planning, siting, and providing generation resources to local customers."[229]

EPA counters, among other things, that it does have authority under Section 111(d) and its longstanding regulations[230] to establish a minimum level of stringency, and that the CPP still allows each state to set particular standards of performance for particular sources:

> Under Section 111(d) ..., the agency promulgates "guidelines" for states to follow when submitting "satisfactory" plans establishing emission standards for existing sources. While it is the states' job to establish such standards, those standards must "reflect[]" the "degree of emission limitation achievable through the application of the [BSER] ... *the*

---

[224] *See* State and Mun. Int. Br. Supp. EPA at 22, 8-25; *see also, e.g.*, Envtl. and Pub. Health Orgs. Int. Br. Supp. EPA at 19-20.

[225] *See* Fmr. EPA Admins. Br. Supp. EPA at 4, 18-31. *See also, e.g.*, Fmr. State Envtl. and Energy Officials Br. Supp. EPA at 18-23.

[226] *See* Cong. Br. Supp. EPA at 4-5; *see generally id.* at 8-14.

[227] *See generally* Pet'rs Br. Core Legal Issues at 74-78; Nev. and Consumers' Res. Br. Supp. Pet'rs at 7-19.

[228] *See* 166 Bus. Ass'ns Br. Supp. Pet'rs at 11.

[229] Pac. Legal Found. Br. Supp. Pet'rs at 5; *see also id.* at 25-26 (citing *FPC v. S. Cal. Edison Co.*, 376 U.S. 205, 215- 16 (1964)).

[230] *See* 40 C.F.R. pt. 60, subpt. B.

> *Administrator* determines has been adequately demonstrated." 42 U.S.C. § 7411(a)(1) (emphasis added [in brief]).[231]

EPA and its supporters also argue, for example, that the CPP generally preserves the existing federal-state division of authority relating to the electrical grid.[232]

Specific Record-Based Challenges

Petitioners jointly submitted two briefs on the merits: the first on "core legal issues" such as those described above, and the second on "procedural and record based issues."[233] As summarized below, many issues within the latter set were also addressed by the amici curiae in support of petitioners. EPA and various of its supporters have largely disputed these procedural and record based challenges.

Achievability and Cost-Benefit Analysis

Petitioners maintain that EPA has not satisfied its legal burden to show that the BSER in the CPP, or its component building blocks, are "adequately demonstrated" or the resulting emission guidelines "achievable" as required under the definition of "standard of performance" in CAA Section 111(a)(1).[234] They also argue that EPA failed to account sufficiently for reliability of the electrical grid or for the need to build new infrastructure, such as transmission lines.[235] In addition, the petitioners oppose EPA's cost-benefit analysis for the Rule as "fundamentally flawed."[236]

As a legal matter, they urge that the potential for states to opt for a multi-state emission credit trading program to meet plan requirements cannot "save" the Rule from these alleged deficiencies in achievability.[237] Petitioners contend that achieving the required emission reductions is an "impossible task" for states.[238] To support this argument, petitioners submitted a letter to the court on September 23, 2016 (four days before the oral argument), citing California's proposed state plan to implement the CPP that would limit trading to states that had plans as stringent as California's requirements.[239] They claim that the proposed California plan that limits its carbon trading program may prevent robust trading systems from developing and shows that EPA has not demonstrated that states such as Montana, Kentucky, North Dakota, and West Virginia will be able to meet their reduction

---

[231] *See* EPA Br. at 73-74.

[232] *See, e.g., id.* at 55-60, 68-76; State and Mun. Int. Br. Supp. EPA at 8-17, 33-35; Fmr. State Envtl. and Energy Officials Br. Supp. EPA at 18-23.

[233] Pet'rs Br. Core Legal Issues; Pet'rs Br. Proc. and R. Issues.

[234] *See generally* Pet'rs Br. Proc. and R. Issues at 17-55.

[235] *See id.* at 38-47.

[236] *See id.* at 69-71.

[237] *See id.* at 49-53.

[238] *See id.* at 53-55.

[239] *See* Letter of State Pet'rs, *West Virginia v. EPA*, No. 15-1363 (D.C. Cir. filed September 23, 2016) (filed pursuant to Rule 28(j) of the Federal Rules of Appellate Procedure).

goals within their own borders.[240] When petitioners raised the proposed California trading program during the oral argument, the judges noted that California's plan was still in the proposal phase and questioned whether these issues should be deferred until states experience such difficulties during planning and implementation.[241]

A number of amici support these general arguments. For example, the brief of the 166 state and local business associations objects that the CPP would result in "devastating economic costs" and "decimate[]" some areas' employment and tax bases by raising costs of operation for American enterprise."[242] Groups representing women, minorities, and seniors, as well as taxpayers, allege that price increases resulting from coal-fired power plant closures and new infrastructure and efficiency requirements would most heavily impact disadvantaged groups.[243] Pedernales Electric Cooperative, which describes itself as "the largest non-profit electric distribution cooperative in the United States," frames the CPP's time frame as "unrealistic" and says that the CPP will have negative impacts on planning, reliability, and security,[244] concerns echoed by the Municipal Electric Authority of Georgia (MEAG).[245] MEAG also describes that it is subject to irrevocable long-term contracts based on specific power plants that will continue to impose payment obligations irrespective of the CPP, and says that adding on new contracts would force its communities to "pay[] twice" for electricity, resulting in negative environmental justice impacts.[246]

EPA responds that it identified an achievable degree of emission limitation by applying the best system, framing its modeling and other analysis as reasonable and its estimates as conservative.[247] EPA spends a substantial portion of its brief walking through its data and approach and working to counter petitioners' factual claims on those points.[248] It also argues, among other things, that it was not required to perform individual plant achievability analyses.[249] EPA states that achieving the emission rates would not require trading, though its analysis in the record demonstrates that trading programs are likely to be established.[250] EPA also contends that it reasonably considered costs, infrastructure, and grid reliability, including specific concerns raised by rural cooperatives and others, and that it reasonably

---

[240] See id. at 2.

[241] See Emily Holden & Debra Kahn, "Clean Power Plan: Inside the Judges' Questions on Carbon Trading," E&E NEWS, September 29, 2016, http://www.eenews.net/stories/1060043590.

[242] See 166 Bus. Ass'ns Br. Supp. Pet'rs at 12-13.

[243] See 60Plus Ass'n et al. Br. Supp. Pet'rs at 4-11.

[244] See Pedernales Br. Supp. Pet'rs at 6, 17; see generally id. at 10-30.

[245] See MEAG Br. Supp. Pet'rs at 13-21.

[246] See id. at viii, 2, 14-18.

[247] *See generally* EPA Br. at 117-64.

[248] *See generally id.*

[249] *See id.* at 142.

[250] *See id.* at 142-46.

calculated and confirmed all of the state-specific goals.[251] EPA cites extensively to its Rule preamble and technical supporting documents in the rulemaking docket.[252]

Intervenors and amici supporting EPA expand on these arguments relating to the achievability of the emission standards and the evidentiary basis for the BSER. Power companies, including cities doing business through their utilities, maintain that EPA appropriately considered the availability of emissions credit trading programs, and deny that the Rule would impair electric reliability in light of the "tremendous flexibility" provided to states and power companies.[253] Environmental and public health organizations also emphasize what they characterize as the "wide array of flexible compliance options,"[254] and explain their view that EPA reasonably applied the statutory factors to determine the degree of emission limitation required.[255] A coalition of wind, solar, and other advanced energy associations spends much of its brief arguing that "[t]he record demonstrates that EPA's determination of the [BSER], and the Building Blocks in particular, was eminently reasonable,"[256] and that EPA reasonably considered other aspects of achievability.[257] Comparable arguments are set forth in briefs submitted by, among others, former state environmental and energy officials.[258] One brief, submitted by a trio of consumers' groups, aims to rebut empirical claims by petitioners and their amici regarding electricity costs, saying that consumer costs would not meaningfully increase and that, rather, the CPP would *reduce* electricity costs by improving efficiency.[259]

State-Specific Objections to Aspects of the CPP

Petitioners contend that the CPP should have been tailored to individual state circumstances. Wisconsin challenges EPA's calculation of its baseline emissions in light of the imminent retirement of a nuclear plant;[260] Arizona and Utah raise issues regarding EPA's accounting for trading between those states and Indian tribes;[261] New Jersey argues

---

[251] *See id.* at 148-74.

[252] *See generally id.* at 117-74 (citing, *inter alia*, Clean Power Plan Final Rule, *supra* footnote 2; EPA, *EPA's Responses to Public Comments on the EPA's Carbon Pollution Emission Guidelines for Existing Stationary Sources: Electric Utility Generating Units* ("Response to Comments") (August 2015), *available at* https://www.regulations.gov/ #!documentDetail; D=EPA-HQ-OAR-2013-0602-37106.

[253] *See* Power Cos. Int. Br. Supp. EPA at 9-14. One utility in particular, Dominion Resources, Inc., in its separate *amici curiae* brief, particularly emphasized the need for flexibility—including emissions trading and averaging approaches— for the CPP to be achievable. *See* Dominion Br. Supp. EPA at 4, 7-17.

[254] *See* Envtl. and Pub. Health Orgs. Int. Br. Supp. EPA at 19-20.

[255] *See id.* at 6-10.

[256] *See* Advanced Energy Ass'ns Int. Br. Supp. EPA at 1.

[257] *See id.* at 4-17.

[258] *See generally* Fmr. State Envtl. and Energy Officials Br. Supp. EPA.

[259] *See generally* Consumer and Ratepayer Orgs. Br. Supp. EPA at 3-31. For more information on the implications of the CPP on electric power system reliability, the costs of electric power to customers, and the future structure of the electric utility industry which could result from implementation of state compliance plans, *see* CRS Report R44265, *EPA's Clean Power Plan: Implications for the Electric Power Sector*, by Richard J. Campbell.

[260] *See* Pet'rs Br. Proc. and R. Issues at 72-73.

[261] *See id.* at 73-75.

that EPA failed to property take into account its deregulation of energy services;[262] North Carolina argues that EPA arbitrarily excluded its emission reductions from consideration;[263] Wyoming charges that EPA ignored its "unique circumstances";[264] and Utah argues that the CPP "would cause particular harm" to that state.[265]

EPA generally counters these arguments one by one in its brief, saying overall that it reasonably calculated all state-specific goals and determined that all states would be able to develop compliant plans.[266] EPA also states that it reasonably determined that pre-2013 generating facilities could not provide emission-rate credits.[267]

During the oral argument, the judges suggested that these issues may be premature and that states will have future opportunities to seek judicial review if the reduction targets prove to be unachievable.[268]

Industry-Specific Objections to Aspects of the CPP

A number of claims pertain primarily to certain industry sectors or subsectors. For example, petitioners claim that in the CPP, EPA "ignores" large parts of the nation's electrical system: "existing renewable energy, nuclear generation that provides approximately 20% of the nation's power with zero emissions, hydroelectric generation that supplies the majority of electricity in many regions of the country, co-generation units, and waste-to-energy facilities with very low carbon footprints."[269] As a result, they say, EPA has "failed to consider an important aspect of the problem."[270] They also object to what they characterize as the Rule's limitations on the use of enhanced oil recovery that also results in associated CO_2 storage,[271] and its lack of different emission guidelines or compliance times for lignite coal-fired power plants.[272]

EPA, in response, cites to the record and argues, among other things, that it adequately explained its treatment of hydropower, nuclear plants, and waste-to-energy facilities.[273] EPA also insists that its limitations and reporting requirements for enhanced oil recovery are reasonable and do not change an oil recovery well's permitting status,[274] and that it reasonably determined that no other subcategories of sources were necessary.[275]

In addition to issues raised in the briefing in *West Virginia v. EPA*, several other petitions for review brought by entities including the National Alliance of Forest Owners,

---

262 *See id.* at 80-82.
263 *See id.* at 82-84.
264 *See id.* at 75-77.
265 *See id.* at 77-80.
266 *See* EPA Br. at 164-74.
267 *See id.* at 164-67.
268 *See* James W. Rubin, *supra* footnote 160; Ellen M. Gilmer, *supra* footnote 141.
269 *See* Pet'rs Br. Proc. and R. Issues at 56-57 (internal footnote omitted).
270 *See id.* at 58 (quoting *Motor Vehicle Mfrs. Ass'n. v. State Farm Mut. Auto. Ins. Co.*, 463 U.S. 29, 43 (1983)).
271 *See id.* at 64.
272 *See id.* at 67.
273 *See, e.g.*, EPA Br. at 164-68.
274 *See id.* at 163.
275 *See id.* at 159-60.

Biogenic CO_2 Coalition, American Forest & Paper Association, and American Wood Council are, at the request of the petitioners and EPA, being held in abeyance pending potential administrative resolution of biogenic GHG emissions issues in the CPP.[276]

Procedural Challenges

Some challengers have disputed the adequacy of certain other procedural aspects of the issuance of the Rule under the CAA and the Administrative Procedure Act.[277] Petitioners charge that the final CPP "could not have been divined from its proposal,"[278] and that "[b]y departing so radically from that proposal, EPA promulgated a Rule on which the public had no opportunity to comment."[279]

EPA maintains that the final CPP is a logical outgrowth of the proposal and comments, and that EPA properly followed all other procedural requirements.[280] EPA, moreover, criticizes petitioners and others for referencing sources and documents that were not made a part of the rulemaking record.[281]

Similar to the state-specific record-based issues discussed above, the judges during oral argument questioned whether the notice issue was ripe for review because petitioners' administrative petitions for review were still pending before EPA.[282] Petitioners argued that EPA had effectively denied their administrative petitions when the agency in its brief argued that it had given adequate notice.[283]

Selected Additional Factual and Policy Issues Briefed by Amici Curiae

Finally, in addition to the many arguments made by the parties to the case, points raised in the briefs of amici curiae expand on the parties' arguments and bring other issues, perspectives, and facts to the court's attention. Again, this chapter does not aim to provide a comprehensive or representative preview of the many legal and factual claims in the CPP

---

[276] Order, *West Virginia v. EPA* (D.C. Cir. January 21, 2016) (holding in abeyance *Nat'l Alliance of Forest Owners v. EPA*, No. 15-1478; *Biogenic CO_2 Coalition v. EPA*, No. 15-1479; and Am. Forest & Paper Ass'n, Inc. and *Am. Wood Council v. EPA*, No. 15-1485). For more information on the potential interactions between the CPP and biomass energy, see CRS Report R41440, *Biopower: Background and Federal Support*, by Kelsi Bracmort, and CRS Report R41603, *Is Biopower Carbon Neutral?*, by Kelsi Bracmort.

[277] *See, e.g.*, Pet'rs Br. Proc. and R. Issues at 11, 13-15 (citing CAA §307(d)(3), 42 U.S.C. 7607(d)(3)). In addition, Petitioner Energy & Environment Legal Institute (EELI) attempted to file a "Supplement to Brief of Petitioners on Procedural and Record-Based Issues" arguing that "EPA violated [CAA procedures and] the procedural due process rights of those affected by failing to place in the public docket during the notice and comment period documents showing substantial *ex parte* contacts which formed the basis of the agency action," based on certain communications between EPA personnel and environmental groups. EELI Br. at 1, 2-5. EPA objected, and the court denied EELI's motion for leave to file the separate supplement. Order, *West Virginia v. EPA* (D.C. Cir. March 21, 2016) (per curiam).

[278] *See* Pet'rs Br. Proc. and R. Issues at 11, 16.

[279] See id. at 11.

[280] *See* EPA Br. at 29, 107-16. For more information on similarities and differences between the proposal and the final CPP, *see* CRS Report R44145, *EPA's Clean Power Plan: Highlights of the Final Rule*, by Jonathan L. Ramseur and James E. McCarthy.

[281] *See, e.g.*, EPA Br. at 40, 137-38, 157-58 (citing CAA Section 307, 42 U.S.C. §7607(d) (review limited to record)).

[282] *See* James W. Rubin, *supra* footnote 160; Ellen M. Gilmer, *supra* footnote 141.

[283] *Id.*

litigation. The many points raised by amici include, but are not limited to, the following points, which—like those previously discussed—have been highly condensed from their original forms.[284]

Supporting Petitioners

- The Pacific Legal Foundation and allies claim that EPA failed to make the required endangerment finding under CAA Section 111 and that EPA could not rely on the endangerment finding that it made in 2009 in the context of motor vehicles.[285] They also insist that if EPA is to regulate GHGs, it may only do so through NAAQS under CAA Section 108, "the regulatory path Congress prescribed for air pollutants in the 'ambient air' emitted from 'numerous or diverse' sources."[286]
- Groups representing seniors, minorities, and women argue, among other things, that "EPA's plans to incentivize investment in low-income communities will do nothing to help those facing immediate increases in electricity rates."[287]
- A group of 13 scientists submitted a brief contending that the lines of evidence cited by EPA in support of its scientific conclusions on the dangers of GHGs and climate change have been "definitively invalidated by real world empirical temperature data,"[288] and that EPA's Social Cost of Carbon analysis is "nonsense that no rational person would use for public policy."[289]

Supporting Respondents

- Former officials Madeleine K. Albright, Leon E. Panetta, and William J. Burns[290] assert that the CPP is "integral to continued U.S. leadership in the fight against climate change," having inspired other countries' commitments to emission reductions,[291] and that "global warming is a national security issue."[292] A brief of the Union of Concerned Scientists argues that the CPP "plays a key role in the worldwide implementation of the breakthrough Paris Agreement."[293]

---

[284] As noted above, this report does not include arguments in petitioners' and their supporters' reply briefs.
[285] *See* Pac. Legal Found. et al. Br. Supp. Pet'rs at 20-24.
[286] *See id.* at 10-15.
[287] *See* 60Plus Ass'n et al. Br. Supp. Pet'rs at 14-16.
[288] *See* Scientists Br. Supp. Pet'rs at 3.
[289] *See id.* at 2.
[290] Madeleine K. Albright served as the Secretary of State from 1997 to 2001 and as U.S. Permanent Representative to the United Nations from 1993 to 1997. Leon E. Panetta served, *inter alia*, as Secretary of Defense for the United States from 2011 to 2013 and director of the Central Intelligence Agency from 2009 to 2011. Career Ambassador William J. Burns served, *inter alia*, as Deputy Secretary of State from 2011 to 2014.
[291] *See* Fmr. Dep't of State and Def. Officials Br. Supp. EPA at 10-11.
[292] *See id.* at 11-14.
[293] *See* UCS Br. Supp. EPA at 14, 14-27.

- Technology companies Amazon.com, Inc., Google Inc., Microsoft Corp., and Apple Inc. (for which former EPA Administrator Lisa P. Jackson is now Vice President of Environmental Initiatives)[294] submitted a brief arguing that the CPP "will help Tech Amici—and countless other companies—power their operations in ways consistent with their environmental commitments and business needs."[295] Several other consumer brand companies also highlight their environmental commitments and state that they would face "economic and social disruptions as a direct result of inaction on regulating power plant emissions."[296]
- Forty-one Christian and Jewish faith groups provided a brief asserting "a moral imperative to protect the Earth and all its inhabitants from a climate crisis of our own making."[297]
- A group of 20 scientists submitted a brief regarding the science of climate change, its impacts (particularly in the United States), and the contribution by combustion of fossil fuels.[298] A coalition of public health groups filed another brief arguing that climate change, "caused by utility sector carbon emissions, has adverse human health impacts," especially on vulnerable populations, in light of impacts on heat, ozone, particulate matter, pollen, and microbial hazards.[299]

Taken together, the briefs in the CPP litigation touch on not only legal and technical issues under CAA Section 111 and administrative law principles, but also broader policy debates regarding the environment, the economy, and governance.

Next Steps in *West Virginia v. EPA*

As noted above, oral argument was held on September 27, 2016.[300] As a practical matter, the court may take some months after oral argument to issue a decision.

Once the D.C. Circuit issues a judgment, a dissatisfied party may seek Supreme Court review.[301] Because of the high stakes of the case and because whichever side is dissatisfied with the result is likely to appeal (and indeed, parties from both sides could file cross-appeals of different aspects of the decision), the case is widely considered a near certainty to reach the Supreme Court, most likely in 2017 or 2018.[302] The D.C. Circuit's decision to hear the

---

294 Apple, Press Info, Apple Leadership: Lisa Jackson, http://www.apple.com/pr/bios/lisa-jackson.html (last visited April 13, 2016).
295 *See* Tech. Br. Supp. EPA at 3; *see generally id.*
296 *See* Adobe, Mars, Ikea, BCBS Br. Supp. EPA at 3, 6; *see generally id.* at 16-24.
297 *See* Faith Grps. Br. Supp. EPA at 2.
298 *See generally* Climate Scientists Br. Supp. EPA.
299 *See* Med. and Pub. Health Groups Br. Supp. EPA at 4; *see generally id.*
300 Order at 2, *West Virginia v. EPA*, No. 15-1363 (D.C. Cir. May 16, 2016).
301 *See Rules of the Supreme Court of the United States*, R. 10-14 (2013).
302 *See, e.g.*, Sidley Austin LLP, White Paper, EFFECT OF SUPREME COURT STAY ON CLEAN POWER PLAN DEADLINES (2016), *available at* http://www.chamberlitigation.com/sites/ default/files/scotus/files/2016/

case en banc in the first instance could potentially streamline the litigation and hasten Supreme Court review somewhat, relative to the initial and ordinary schedule in which a case must first be decided by a three-judge panel before any petitions may be filed for rehearing of the panel decision en banc.

North Dakota v. EPA: Section 111(b) Litigation

In addition to the direct legal challenge to the CPP rule for CO_2 from existing power plants under CAA Section 111(d), 25 states—led by North Dakota and West Virginia—have filed petitions in the D.C. Circuit challenging EPA's final NSPS rule for CO_2 from new, modified, and reconstructed power plants under CAA Section 111(b), which it calls the "Carbon Pollution Standards."[303] The states have been joined by other petitioners including a labor union, a rural electric cooperatives association, several other fossil-fuel-related companies and utilities, and several industry and trade groups; most of the petitioners overlap with those who also filed challenges to the CPP, although there are somewhat fewer petitioners challenging the NSPS. The petitions have been consolidated under the case caption *North Dakota v. EPA*.[304] Most of the states and a number of the nonprofit organizations that intervened in support of EPA in the CPP case also intervened in the NSPS challenge in support of EPA.[305]

In the Section 111(b) litigation, one of the primary issues is EPA's establishment of standards of performance based on technologies including carbon capture and sequestration/storage (CCS).[306] Natural gas plants and modified coal plants can reach the final NSPSs with efficient generation technology, but new coal plants would need to

White%20Paper%20on%20Impact%20of%20Stay%20on%20CPP%20Deadlines.pdf (predicting time frame for Clean Power Plan litigation after Supreme Court stay and arguing for tolling of all CPP deadlines for the length of the duration of the litigation if the Rule is upheld); James W. Rubin, *Clean Power Plan Finally Has Its Day In Court—Now What?* LAW360, October 1, 2016, https://www.law360.com/environmental/articles/847138/clean-power-plan-finallyhas-its-day-in-court-now-what- (projecting that the case may "not be heard [by the Supreme Court] until fall of 2017 or even later. Legal uncertainty thus could continue into 2018.").

303 *See generally* docket for *North Dakota v. EPA*, No. 15-1381 (D.C. Cir. filed October 23, 2015); Standards of Performance for Greenhouse Gas Emissions From New, Modified, and Reconstructed Stationary Sources: Electric Utility Generating Units; Final Rule, 80 Fed. Reg. 64,510 (October 23, 2015) (hereinafter "NSPS Final Rule"). As with the CPP litigation, several websites have been maintaining compilations of major filings and orders in the case, including the Chamber of Commerce of the United States, a petitioner, at its *U.S. Chamber Litigation Center*, http://www.chamberlitigation.com/chamber-commerce-et-al-v-epa-nsps-rule; and the Environmental Defense Fund, an intervenor in support of EPA, at its *Clean Power Plan Case Resources* website at https://www.edf.org/climate/cleanpower-plan-case-resources (tab for "111(d)"). Colorado and New Jersey did not join the coalition of states challenging the NSPS rule.

304 *See generally* docket for *North Dakota v. EPA*, *supra* footnote 303.

305 *See id.*

306 *See, e.g.*, Chamber of Commerce of the United States of America, et al., Pet'rs Prelim. Stmt. of Issues to be Raised, *North Dakota v. EPA*, Nos. 15-1381, 15-1469 (D.C. Cir. filed January 25, 2016) ("Chamber of Commerce Stmt. of Issues"); Murray Energy Corp., Stmt. of Issues to be Raised, *North Dakota v. EPA*, Nos. 15-1381, 15-1396 (D.C. Cir. filed December 2, 2015) ("Murray Energy Stmt. of Issues").

implement partial CCS.[307] Critics of the NSPSs for power plants say, for example, that CCS technology is not yet commercially available nor fully technically feasible, and therefore that it is not "adequately demonstrated" or the "best system" under Section 111(b).[308] They also argue that EPA improperly relied on separate demonstrations of individual components of the technology, and that the NSPSs are otherwise arbitrary, capricious, an abuse of discretion, contrary to law, or unconstitutional.[309] EPA, in the Section 111(b) final rule's preamble, argued that its Carbon Pollution Standards were reasonable and lawful.[310] EPA provided rationales for basing the standard for new coal plants on partial CCS: It explained technical configurations and operational flexibilities that may be available; worked through its analyses of feasibility, cost, and other criteria; and discussed "alternative compliance options that new source project developers can elect to use, instead of ... partial CCS, to meet the final standard of performance."[311]

In addition to the D.C. Circuit litigation, petitions were filed before EPA seeking administrative reconsideration of the NSPS rule; EPA denied the petitions in May 2016. The D.C. Circuit suspended the briefing schedule in the *North Dakota v. EPA* case in June 2016 to allow the court to consolidate new lawsuits filed challenging EPA's denial of the petitions. On August 30, 2016, the court issued a revised briefing schedule to begin on October 13, 2016, with opening briefs and conclude by February 6, 2017.[312] The court has scheduled oral arguments for April 17, 2017.

As noted above, the finalization of NSPSs for new air pollutant sources under Section 111(b) of the CAA is a prerequisite for the use of authority under Section 111(d) to regulate existing sources, so this litigation could threaten EPA's basis for the CPP.[313] Thus, regardless of the outcome of *West Virginia v. EPA*, the litigation in *North Dakota v. EPA* potentially could impact the CPP as well.

---

[307] *See* NSPS Final Rule, *supra* footnote 303, 80 Fed. Reg. at 64,513-14, 64,543-97.

[308] *See generally, e.g.*, Chamber of Commerce Stmt. of Issues; Murray Energy Stmt. of Issues.

[309] *See, e.g.*, Chamber of Commerce Stmt. of Issues; Murray Energy Stmt. of Issues; *see also, e.g.*, Energy & Env't Legal Inst., Pet'r Stmt. of Issues to Be Raised, at 3, *North Dakota v. EPA*, Nos. 15-1381, 15-1397 (D.C. Cir. filed December 2, 2015) (raising issue that "the Final Rule creates an unconstitutional taking of property interests that can be avoided by an interpretation that is more consistent with the plain text of the rule and more consistent with past practice").

[310] *See generally* NSPS Final Rule, *supra* footnote 303.

[311] *Id.*, 80 Fed. Reg. at 64547-48; *see generally id.* at 64,547-97.

[312] *See* Order, *North Dakota v. EPA*, No. 15-1381 (D.C. Cir. August 30, 2016).

[313] *See supra*, "Clean Air Act Section 111."

In: The Clean Air Act and The Clean Power Plan
Editor: Claire Ferguson
ISBN: 978-1-53614-825-1

Chapter 10

EPA PROPOSES TO REPEAL THE CLEAN POWER PLAN*

Linda Tsang

UPDATE: On December 28, 2017, the U.S. Environmental Protection Agency (EPA) published an Advanced Notice of Proposed Rulemaking (ANPRM) seeking information on a potential replacement to the Clean Power Plan (CPP)—the emissions guidelines for greenhouse gases (GHGs) from existing power plants that the agency proposed to repeal in October 2017. An ANPRM allows an agency to gather information for potential rulemaking but does not bind an agency to future action. As explained in the ANPRM, EPA is "considering" proposing a rule to replace the CPP and seeks comments on the (1) "roles, responsibilities, and limitations of the federal government, state governments, and regulated entities in developing and implementing" the new GHG emission guidelines; (2) the "best system of emission reduction" consistent with EPA's new legal interpretation set forth in the proposed repeal of the CPP; and (3) approaches to harmonize and streamline the process and applicability of Clean Air Act permitting for existing power plants regulated under the new emission guidelines. EPA will accept comments on the ANPRM until February 26, 2018. This Legal Sidebar post originally appeared November 13, 2017.

On October 10, 2017, the Environmental Protection Agency (EPA) issued a proposal to repeal the Obama Administration's 2015 rule, "Carbon Pollution Emission Guidelines for Existing Stationary Sources: Electric Utility Generating Units (EGUs)" (commonly referred to as the Clean Power Plan (CPP) rule). The CPP has not gone into effect because the Supreme Court in 2016 stayed the implementation of the rule until the lawsuit challenging

* This is an edited, reformatted and augmented version of a Congressional Research Service report, Publication No. LSB10016, dated January 4, 2018.

its legality is resolved. Upon its review of the CPP and its 2015 legal justification, EPA has now determined that the CPP exceeds its statutory authority based on a change in the agency's legal interpretation of Section 111 of the Clean Air Act (CAA). The proposal formally starts a potentially lengthy process to repeal the CPP and raises questions about whether EPA will replace the CPP with another rule targeting CO_2 emissions from existing power plants and how the repeal will affect existing legal challenges to the CPP. This Sidebar will explore these and other questions and the next steps in repealing the CPP.

What Is the Legal Basis for EPA's Proposed Repeal of the CPP?

EPA is proposing to repeal the CPP after reviewing the rule and its 2015 legal justification, in accordance with the Trump Administration's Executive Order 13783, "Promoting Energy Independence and Economic Growth." Upon review of the CPP, EPA determined that its 2015 legal interpretation of the "best system of emission reduction" (BSER) for CO_2 from existing power plants exceeds its statutory authority under CAA Section 111(d).

As background, Section 111 directs EPA to list categories of stationary sources that cause or contribute significantly to "air pollution which may reasonably be anticipated to endanger public health or welfare." Once EPA lists a source category such as fossil fuel-fired EGUs (power plants), Section 111(b) requires EPA to establish "standards of performance" for new and modified sources (known as NSPSs) within a listed source category. Under Section 111(a), a "standard of performance" is defined as "a standard for emissions of air pollutants which reflects the degree of emission limitation achievable through the application of the best system of emission reduction [BSER]." Once NSPSs are issued under Section 111(b) for *new or modified* sources in that category, EPA is then required to establish "emission guidelines" for states to set a "standard of performance" for *existing* sources under Section 111(d).

In 2015, EPA finalized both NSPSs for *new* or *modified* power plants under Section 111(b) and emission guidelines for *existing* power plants (the CPP) under Section 111(d). For the CPP, EPA based the BSER for existing power plants on three "building blocks": (1) improving efficiency at affected power plants, (2) shifting generation from higher-emitting coal units to lower-emitting natural gas combined cycle units, and (3) shifting generation from fossil fuel units to renewable energy generation. The agency then used the BSER to set a CO_2 emission "standard of performance" for existing power plants.

EPA's proposed repeal is based on a different legal interpretation of the BSER from its 2015 interpretation. As explained in the proposal, a broader interpretation of the BSER necessarily yields a "greater universe" of measures that EPA could consider in setting emission standards, and, conversely, a narrower reading limits the measures the agency can consider for the affected sources. EPA proposes to limit the interpretation of the BSER to emission reduction measures that can be "*applied to or at* the individual source, . . . rather than measures that the source's owner or operator *can implement on behalf of* the source at another location." In other words, EPA proposes that it has authority to regulate only emission control technologies at individual power plants and not broader measures that can be implemented outside of the physical location of the individual power plant. EPA explains in the proposed repeal that physical and operational efficiency improvements could be applied directly to an affected source, while "beyond-the-source" measures (commonly referred to as "outside the fenceline" measures) such as generation shifting and renewable energy capacity increases could be applied only by owners and operators at a different location. Because the BSER in the 2015 final rule was based, in part, on "beyond-the-source" measures, EPA concludes that the CPP exceeded its authority under the CAA and proposes that it be repealed.

In both the 2015 CPP rule and the proposed repeal, EPA cites to CAA legislative history, prior agency practice, jurisprudence, and policy concerns to justify its interpretation of the BSER. However, the proposed repeal focuses on different aspects and perspectives of those materials to support a source-oriented and narrower interpretation of the BSER. For example, EPA argues that the BSER should be interpreted similar to other technology-based CAA standards such as the "maximum achievable control technology standards" for hazardous air pollutants that apply physical and operational measures directly at the source. While acknowledging that a source's owner or operator implements the measures, EPA explains that its 2017 interpretation focuses on what can be applied directly to the source "from the perspective of the source and not its owner or operator."

This proposed interpretation of the BSER directly contrasts with the interpretation set forth in 2015 that focused on the measures that the owner or operator could implement. EPA's 2015 interpretation concluded that "system" in the BSER "is capacious enough to include actions taken by the owner/operator of a stationary source designed to reduce emissions from that affected source, including actions that may occur off-site and actions that a third party takes pursuant to a commercial relationship with the owner/operator." EPA reasoned in the CPP that the electricity "system" should be viewed as a group of interdependent and integrated individual power plants when determining the BSER.

WHAT ARE THE NEXT STEPS IN THE REPEAL?

In order to repeal a rule, an agency must generally follow the same notice and comment rulemaking procedures that were required to issue the rule in the first place. Here, because the CPP went through notice and comment rulemaking, EPA is following the same process to repeal the rule. EPA will accept comments on the proposal received by January 16, 2018.

EPA will hold a hearing on the proposal in Charleston, West Virginia on November 28 and 29, 2017. Many stakeholders likely will submit comments on the proposed repeal. In 2015, the proposed CPP received more than 4.3 million public comments, the most ever for an EPA rule. Under the CAA Section 307(d), the agency may issue a final rule repealing the CPP once it has considered any comments received on the proposal.

Can the Repeal Be Challenged in Court?

Under CAA Section 307(b), if EPA finalizes the repeal of the CPP, a petition for review of the final repeal can be filed with the U.S. Court of Appeals for the District of Columbia Circuit (D.C. Circuit) within 60 days from the date the repeal is published in the *Federal Register*. The court may reverse an agency action that the court finds to be:

- arbitrary, capricious, an abuse of discretion, or otherwise not in accordance with law;
- contrary to a constitutional right, power, privilege, or immunity;
- in excess of statutory jurisdiction, authority, or limitations, or short of statutory right; or
- an arbitrary and capricious failure to observe legal procedures.

Petitioners of the repeal may also ask the court to stay (i.e., pause) the repeal, similar to the petitioners' request to stay the implementation and enforcement of the CPP.

What Will Happen to the CPP Litigation in the D.C. Circuit?

The D.C. Circuit granted EPA's request to hold in abeyance (i.e., pause temporarily) the CPP litigation until October 10, 2017, to allow EPA to reconsider the CPP. Because the repeal of the CPP has not been finalized, it is difficult to predict the next steps for the litigation. The court could remand or continue to pause the litigation—actions that the court considered when it first issued the abeyance. EPA recently filed a litigation report on the status of its reconsideration of the CPP, explaining that the court should continue to hold the case in abeyance until the repeal is finalized. EPA could also seek to dismiss the case as moot once the repeal is final.

States and other groups supporting the CPP in the litigation urge the court to issue a decision that could guide any potential replacement of the CPP or potential legal challenges to the repeal. Many of the legal issues raised in the CPP litigation, including the scope of

EPA's authority and the interpretation of the BSER, likely will be central to any future legal challenges to the repeal or replacement rule of the CPP. For example, in the litigation, petitioners claimed that CPP is unlawful because Congress must issue a "clear statement" of authority for an agency action that could have potentially serious economic and political implications. EPA is seeking comment on this issue in the proposed repeal. However, the court may not want (and could potentially be constrained by the prohibition against a court ruling on a case that is moot) to expend judicial resources and time to issue a decision for a regulation that likely will be altered or repealed soon.

Will EPA Issue a Replacement for the CPP to Regulate CO_2 Emissions from Power Plants?

On the same day that EPA issued the proposed repeal, EPA submitted for White House review an advance notice of proposed rulemaking (ANPRM) that considers a replacement for the CPP. An ANPRM allows an agency to gather information for potential rulemaking but does not bind an agency to future action. The proposed repeal indicated that EPA will consider through an ANPRM whether it is appropriate to issue another rule interpreting CAA Section 111(d) to replace the CPP and if so, in what form and scope.

Stakeholders and two members of the D.C. Circuit have indicated that EPA may have a legal obligation to replace the CPP. In the order extending the abeyance of the CPP litigation, Judges Tatel and Millett issued a concurring opinion, noting that EPA has an "obligation" to regulate greenhouse gases (GHGs) from power plants because the agency's 2009 determination that GHGs in the atmosphere endangers public health and welfare "triggered an affirmative statutory obligation" to make such a regulation. To date, EPA has not announced plans to reconsider the 2009 "endangerment finding."

In addition, stakeholders may turn to the courts to determine whether EPA must issue a replacement for the CPP. In 2011, EPA entered into an agreement to settle a lawsuit brought by states and environmental groups in which EPA agreed to set standards for GHG emissions from new and existing fossil fuel-fired power plants under CAA Section 111. If EPA does not issue a replacement for the CPP, the states and groups party to the settlement could seek judicial relief to enforce the terms of the agreement, which likely would involve a new lawsuit that would review whether EPA breached the agreement by not complying with its terms.

Without the CPP or a replacement, the power industry likely will still undergo change as the result of market forces and the need to update or replace older generation facilities, issues that are discussed in this chapter.

What Will Happen to the GHG Emission Standards for New and Modified Power Plants?

Pursuant to Executive Order 13783, EPA is currently reviewing the New Source Performance Standards (NSPSs) for GHG emissions from new and modified power plants issued under CAA Section 111(b). The NSPSs for power plants is currently in effect and applies to newly constructed or modified fossil fuel-fired power plants. The D.C. Circuit issued an order to pause the litigation challenging the NSPSs to allow EPA to reconsider the standards. Because EPA must issue Section 111(b) NSPSs before it may issue emission guidelines under Section 111(d) for existing sources, a repeal of the NSPSs could threaten EPA's basis for the CPP or any replacement. Because EPA has not completed its review, it is unclear what actions, if any, EPA will take with respect to the NSPSs for new and modified power plants.

What Actions Might Congress Take?

Congress has taken an active interest in the fate of the CPP since it was proposed in 2014 and subsequently challenged in court. For instance, Members of Congress have filed amici curiae briefs on both sides of the CPP litigation. A brief opposing the CPP argues, among other things, that EPA "usurped the role of Congress" through the CPP's "expansive regulatory requirements." A brief in support of the CPP argues, among other things, that Congress conferred "broad authority" on EPA, and that the CPP is "consistent with the text, structure, and history" of the CAA. Some Members of Congress may consider submitting comments on the proposed repeal and ANPRM or propose legislation that clarifies the scope of EPA's authority under CAA Section 111 and the definition of the BSER.

In: The Clean Air Act and The Clean Power Plan
Editor: Claire Ferguson
ISBN: 978-1-53614-825-1

Chapter 11

U.S. CARBON DIOXIDE EMISSIONS TRENDS AND PROJECTIONS: ROLE OF THE CLEAN POWER PLAN AND OTHER FACTORS*

Jonathan L. Ramseur

SUMMARY[1]

Recent international negotiations and domestic policy developments have generated interest in current and projected U.S. greenhouse gas (GHG) emission levels. GHG emissions are generated throughout the United States from millions of discrete sources. Of the GHG source categories, carbon dioxide (CO_2) emissions from fossil fuel combustion account for the largest percentage (77%) of total U.S. GHG emissions. The electric power sector contributes the second largest percentage (35%) of CO_2 emissions from fossil fuel combustion (one percentage point behind the transportation sector).

In December 2015, delegations from 195 nations, including the United States, adopted an agreement in Paris that creates an international structure for nations to pledge to abate their GHG emissions, adapt to climate change, and cooperate to achieve these ends, including financial and other support. Pursuant to that agreement, the United States pledged (in 2015) to reduce GHG emissions by 26-28% by 2025 compared to 2005 levels. At the date of this chapter, U.S. involvement in the Paris Agreement remains uncertain. However, some recent reports indicate that President Trump is expected to withdraw from the agreement.

U.S. GHG levels in 2015 were 11% below 2005 levels. Whether the United States achieves its goals would likely depend, to some degree, on CO_2 emissions from power plants. In 2015, under President Obama, the Environmental Protection Agency (EPA) promulgated standards for CO_2 emissions from existing electric power plants. The rule,

---

* This is an edited, reformatted and augmented version of a Congressional Research Service report, Publication No. R44451, dated May 31, 2017.

[1] Amber Hope Wilhelm, a Visual Information Specialist, contributed to this report.

known as the Clean Power Plan (CPP), is the subject of ongoing litigation involving a number of entities. On February 9, 2016, the Supreme Court stayed the rule for the duration of the litigation.

Multiple factors generally impact CO_2 emission levels from the electric power sector. Some factors are listed below in no particular order:

- Economic growth/recession,
- Relative prices of energy sources for electricity—particularly natural gas and renewable energy sources,
- Electricity generation portfolio (i.e., the ratio of electricity generation from coal, natural gas, and renewable energy sources),
- National and/or state policy developments (e.g., CPP implementation), and
- Demand-side efficiency improvements (e.g., commercial and residential electricity use).

Recent changes in the electric power sector may be informative. Between 1975 and 2010, electricity generation and CO_2 emissions from the electric power sector generally increased. However, in 2010, their courses diverged. While electricity generation remained relatively flat after 2010, CO_2 emissions from the electric power sector decreased. Thus in 2016, electricity generation was essentially equivalent to generation in 2005, while CO_2 emissions were 25% below 2005 levels.

Recent changes in the U.S. electricity generation portfolio played a key role in the CO_2 emission decrease. The electricity portfolio affects CO_2 emission levels because different sources of electricity generation produce different rates of CO_2 emissions per unit of electricity (zero in the case of some renewables). For example, between 2005 and 2016:

- Coal's contribution to total electricity generation decreased from 50% to 30%;
- Natural gas's contribution to total electricity generation increased from 19% to 34%; and
- Renewable energy's contribution to total electricity generation increased from 2% to 8%.

Accurately forecasting future CO_2 emission levels is a complex and challenging endeavor. A comparison of actual CO_2 emissions (from energy use) between 1990 and 2017 with selected emission projections illustrates this difficulty. In general, actual emissions have remained well below projections.

As the future of the CPP is uncertain, some have questioned whether existing policies and trends in electricity generation would continue to lower CO_2 emissions. Modeling results indicate that CO_2 emissions in the electricity sector are expected to continue declining. However, modeling results indicate that the declines would be substantially greater if the CPP were implemented.

INTRODUCTION

U.S. greenhouse gas (GHG) emission levels, particularly from carbon dioxide (CO_2), remain a topic of interest among policymakers and stakeholders.[2] Recent international

[2] The primary GHGs associated with human activity (and estimated by EPA in its annual inventories) include carbon dioxide, methane, nitrous oxide, sulfur hexafluoride, chlorofluorocarbons, hydrofluorocarbons, and perfluorocarbons.

negotiations and domestic policy developments have generated attention to current and projected U.S. GHG emission levels. An understanding of GHG emission source data and the underlying factors that affect emission levels might help inform the discussion among policymakers regarding GHG emission mitigation.

In December 2015, delegations from 195 nations, including the United States, adopted an agreement in Paris that creates an international structure for nations to pledge to abate their GHG emissions, adapt to climate change, and cooperate to achieve these ends, including financial and other support.[3] Pursuant to that agreement, the United States pledged (in 2015) to reduce GHG emissions by 26-28% by 2025 compared to 2005 levels.[4] This pledge supplemented a previous Obama Administration commitment to reduce U.S. GHG emissions by 17% below 2005 levels by 2020.[5]

At the date of this chapter, U.S. involvement in the Paris Agreement remains uncertain.[6] However, some recent reports indicate that President Trump is expected to withdraw from the agreement.[7]

Whether the United States ultimately achieves the 2020 and 2025 targets will likely depend, to some degree, on GHG emission levels, particularly CO_2 emissions, from electric power plants— one of the largest sources of U.S. emissions. During the Obama Administration, the Environmental Protection Agency (EPA) promulgated standards for CO_2 emissions from existing fossil-fuel-fired electric power plants.[8] EPA cited Section 111(d) of the Clean Air Act (CAA) as the authority to issue its final rule.[9] The rule, known as the Clean Power Plan (CPP), appeared in the *Federal Register* on October 23, 2015.[10]

The fate of the CPP is uncertain. First, the rule is the subject of ongoing litigation. A number of states and other entities have challenged the rule, while other states and entities have intervened in support of the rule. On February 9, 2016, the Supreme Court stayed the rule for the duration of the litigation. The rule therefore currently lacks enforceability or legal effect, and if the rule is ultimately upheld, at least some of the deadlines would likely be delayed.[11]

Second, President Trump issued an Executive Order on March 28, 2017, that directs EPA to review the CPP (and other rulemakings) and "as soon as practicable, suspend,

---

[3] See CRS Insight IN10413, *Climate Change Paris Agreement Opens for Signature*, by Jane A. Leggett.

[4] U.S. Government, "U.S. Cover Note, INDC and Accompanying Information," March 31, 2015, http://www4.unfccc.int/submissions/indc/Submission%20Pages/submissions.aspx.

[5] Executive Office of the President, "The President's Climate Action Plan," June 2013, http://www.whitehouse.gov/sites/default/files/image/president27sclimateactionplan.pdf.

[6] For more information on the withdrawal process, see CRS Report R44761, *Withdrawal from International Agreements: Legal Framework, the Paris Agreement, and the Iran Nuclear Agreement*, by Stephen P. Mulligan.

[7] See Michael Shear and Coral Davenport, "Trump Poised to Pull U.S. from Paris Climate Accord," *New York Times*, May 31, 2017.

[8] See CRS Report R44341, *EPA's Clean Power Plan for Existing Power Plants: Frequently Asked Questions*, by James E. McCarthy et al.

[9] 42 U.S.C. §7411(d).

[10] EPA, "Carbon Pollution Emission Guidelines for Existing Stationary Sources: Electric Utility Generating Units," Final Rule, 80 *Federal Register* 64661, October 23, 2015.

[11] See CRS Report R44341, *EPA's Clean Power Plan for Existing Power Plants: Frequently Asked Questions*, by James E. McCarthy et al.

revise, or rescind the guidance, or publish for notice and comment proposed rules suspending, revising, or rescinding those rules."[12] In April, EPA began its review of the CPP and related power plant rules and may consider various options to revise or repeal the rules.[13]

A question for policymakers is whether U.S. GHG emissions will remain at current levels, decrease to meet the 2020 and 2025 goals, or increase toward former (or even higher) levels. Multiple factors—including economics, technology, and climate policies—will likely play a role in future GHG emission levels.

This chapter examines recent trends in U.S. GHG emissions, particularly CO_2 emissions from electricity generation, and the factors that impact emission levels in that sector. In addition, this chapter examines the degree to which CPP implementation (or lack thereof) may impact CO_2 emission levels from electric power plants.

The first section provides an overview of various sources of GHG emissions in the United States. This includes an overview of CO_2 emissions from fossil fuel combustion and a closer look at CO_2 emissions from electricity generation, which account for the second largest percentage of CO_2 emission from fossil fuel combustion (1% percentage point behind the transportation sector). The second section examines projections of CO_2 emissions in the electric power sector, with a particular focus on the role of the CPP and other factors. The final section highlights the challenges in making CO_2 emission projections with a comparison of actual CO_2 emissions with prior emission forecasts.

Emissions Data in This Chapter

This chapter uses GHG emissions data from two different sources: EPA and the Energy Information Administration (EIA). Estimates of total GHG emissions ("economy-wide") come from EPA's annual GHG emissions inventory. These estimates provide a big-picture view of U.S. GHG emission levels and GHG emission sources, particularly in the context of recent GHG emission reduction goals. EPA released a draft of the most recent version of its inventory in April 2017. This version includes GHG emissions data through 2015. In addition, the CO_2 data in EPA's CPP modeling results come from EPA. EPA released these results in 2015.

Although EPA's Inventory includes CO_2 emissions, this chapter uses CO_2 emissions data from EIA, because EIA's CO_2 emissions data are released on a monthly basis, including annual numbers for 2016. This allows for more recent comparisons of trends in emissions and related topics.

---

[12] Executive Order 13783, "Promoting Energy Independence and Economic Growth," 82 *Federal Register* 16093, March 31, 2017 (signed March 28, 2017).

[13] EPA, "Review of the Clean Power Plan," 82 *Federal Register* 16329, April 4, 2017; EPA, "Review of the Standards of Performance for Greenhouse Gas Emissions From New, Modified, and Reconstructed Stationary Sources: Electric Generating Units," 82 *Federal Register* 16330, April 4, 2017.

U.S. GHG EMISSIONS

Figure 1 illustrates U.S. GHG between 1990 and 2015. As the figure indicates, U.S. GHG emissions increased during most of the years between 1990 and 2007. GHG emissions decreased substantially in 2008 and 2009 as a result of a variety of factors—some economic, some the effect of government policies at all levels. Over the last five years, emissions have fluctuated but have not surpassed 2009 levels. Emissions in 2015 were roughly equivalent to 1994 emission levels.

Figure 1 compares recent U.S. GHG emission levels to the 2020 and 2025 emission goals. Based on 2015 GHG emission levels, the United States is more than halfway to reaching the Administration's 2020 goal (17% below 2005 levels). U.S. GHG levels in 2015 were 11% below 2005 levels.

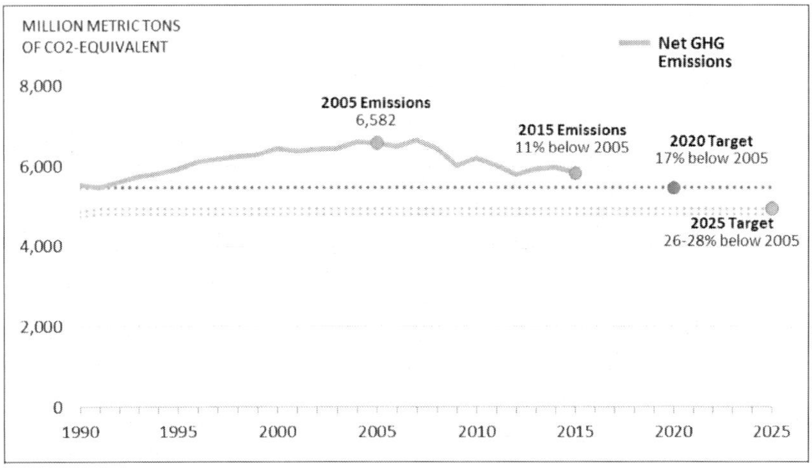

Source: Prepared by CRS; data from EPA, *Inventory of U.S. Greenhouse Gas Emissions and Sinks: 1990–2015*, April 2017, https://www.epa.gov/ghgemissions/inventory-us-greenhouse-gas-emissions-and-sinks-1990-2015.

Notes: Net GHG emissions includes net carbon sequestration from Land Use, Land Use Change, and Forestry. This involves carbon removals from the atmosphere by photosynthesis and storage in vegetation.

Figure 1. U.S. GHG Emissions (Net); Compared to 2020 and 2025 Emission Targets.

Accurately forecasting future emission levels is a complex and challenging endeavor. Consequently, analysts often provide a range of emissions based on different scenarios or assumptions. The Energy Information Administration (EIA) provides annual forecasts of CO_2 emissions in its *Annual Energy Outlook* (AEO) publications. Regarding its various estimates, EIA states the following:

> Projections in the Annual Energy Outlook 2017 (AEO2017) are not predictions of what will happen, but rather modeled projections of what may happen given certain assumptions and methodologies.... The Reference case projection assumes trend

> improvement in known technologies, along with a view of economic and demographic trends reflecting the current central views of leading economic forecasters and demographers. It generally assumes that current laws and regulations affecting the energy sector, including sunset dates for laws that have them, are unchanged throughout the projection period.[14]

Figure 2 compares actual CO_2 emissions between 1990 and 2016 with selected EIA emission projections made in past years. In general, actual emissions have remained well below projections. For example, the AEO from 2002 projected that CO_2 emissions would be almost 6.9 billion metric tons in 2015, about 33% higher than observed emissions. By comparison, the more recent projections (AEO 2014 and AEO 2017) indicated that CO_2 emissions would remain relatively flat or experience modest declines over the next decade.

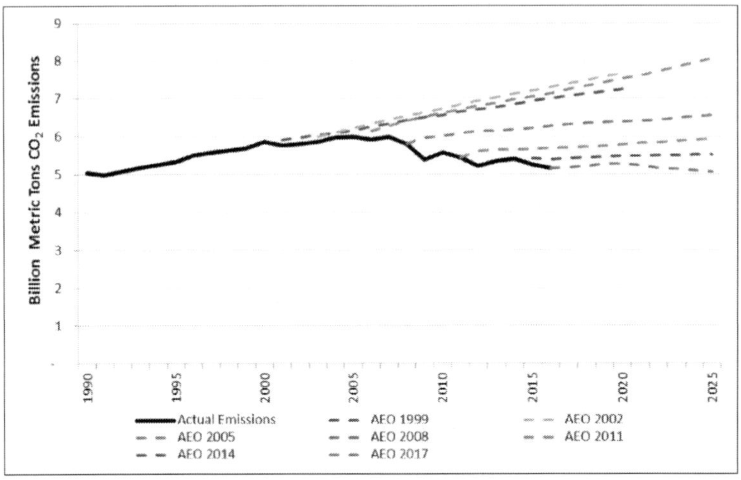

Source: Prepared by CRS; data from EIA *Annual Energy Outlook* publications, http://www.eia.gov.

Notes: EIA publishes annual projections. The above figure includes only projections from every third year since 1999. Some of the earlier projections stopped at 2020.

Figure 2. Actual CO_2 Emissions and Selected Past EIA CO_2 Emission Projections; CO_2 Emissions from Energy Use.

GHG Emission Sources

GHG emissions are generated throughout the United States from millions of discrete sources: power plants, vehicles, households, commercial buildings, agricultural activities (e.g., soils and livestock), and industrial facilities.[15] Figure 3 illustrates the breakdown of U.S. GHG emissions by gas and type of source. The figure indicates that CO_2 from the combustion of fossil fuels — petroleum, coal, and natural gas—accounted for 77% of total

[14] EIA, Annual Energy Outlook 2017, January 2017.

[15] GHG emissions are also released through a variety of natural processes such as methane emissions from wetlands. This report focuses on human-related (anthropogenic) GHG emissions.

U.S. GHG emissions in 2015. Recent legislative proposals that would address climate change have primarily focused on CO_2 emissions from fossil fuel combustion.

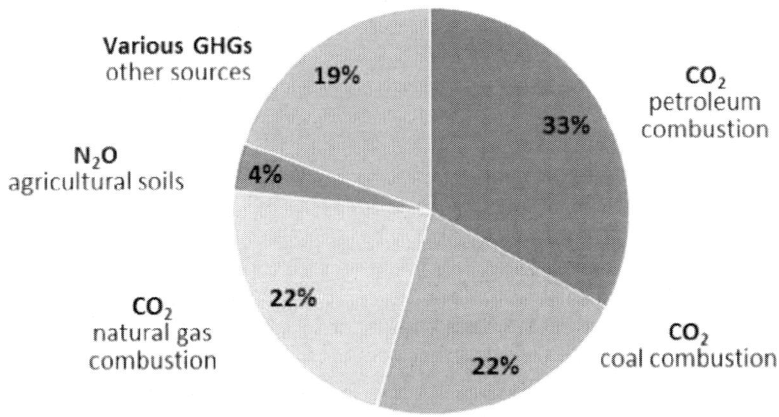

Source: Prepared by CRS; data from EPA, *Inventory of U.S. Greenhouse Gas Emissions and Sinks: 1990–2015*, April 2017, https://www.epa.gov/ghgemissions/inventory-us-greenhouse-gas-emissions-and-sinks-1990-2015.

Notes: N_2O is nitrous oxide. The "Various GHGs – other sources" include the following:

Methane (CH_4) from livestock (3%); hydrofluorocarbons from the substitution of ozone-depleting substances (3%); CO_2 from non-energy fuel uses (2%); CH_4 from natural gas systems (2%), CH_4 from landfills (2%); CO_2 from iron and steel production (1%); CH_4 from coal mines (1%); and CH_4 from manure management (1%). Multiple smaller sources account for the remaining 6%. These percentages may not add up precisely due to rounding.

GHG emissions are typically measured in tons of CO_2-equivalent. This term of measure is used because GHGs vary by global warming potential (GWP). GWP is an index of how much a GHG may contribute to global warming over a period of time, typically 100 years. GWPs are used to compare gases to CO_2, which has a GWP of 1. For example, in EPA's GHG Inventory, methane's GWP is 25, and thus a ton of methane is 25 times more potent a GHG than a ton of CO_2 (over a 100-year lifetime).

Figure 3. U.S GHG Emissions by Source; 2015 Data Measured in Metric Tons of CO_2-equivalent.

CO_2 Emissions from Fossil Fuel Combustion

Figure 4 illustrates the U.S. CO_2 emission contributions by sector from the combustion of fossil fuels. The electric power sector contributes the second largest percentage (35%) of CO_2 emissions from fossil fuel combustion (one percentage point behind the transportation sector). Within the electricity sector, the residential and commercial sectors account for 14% and 13%, respectively, of fossil fuel combustion CO_2 emissions and the industrial sector accounts for 9% of fossil fuel combustion CO_2 emissions (Figure 4).[16]

[16] CO2 emissions related to electricity use in the transportation sector account for less than 1% of CO2 emissions from total electricity generation.

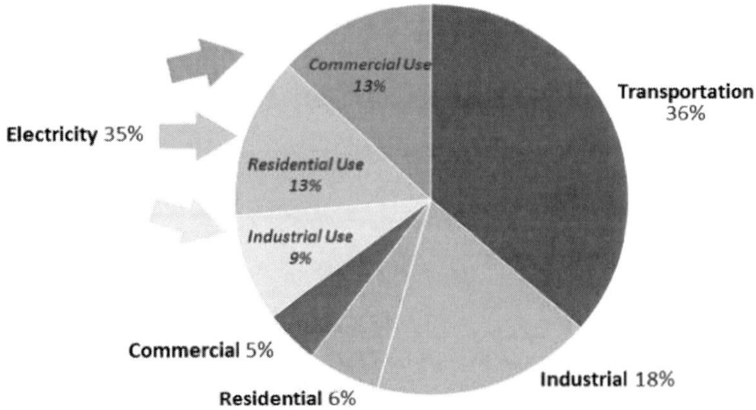

Source: Prepared by CRS; data from EIA, *Monthly Energy Review*, Tables 12.2-12.6, https://www.eia.gov/ totalenergy/data/monthly/.

Notes: CO_2 emissions related to electricity use in the transportation sector account for less than 1% of CO_2 emissions from total electricity generation. These emissions are not included in the above figure. In addition, the above chart does not include CO_2 emissions from the U.S. territories, which account for less than 1% of CO_2 emissions from energy consumption.

The data in this figure do not include emissions associated with various processes that may be generated prior to combustion (e.g., fugitive CH4 emissions from natural gas production). For more details on this issue, see CRS Report R44090, *Life-Cycle Greenhouse Gas Assessment of Coal and Natural Gas in the Power Sector*, by Richard K. Lattanzio.

Figure 4. U.S. CO_2 Emissions from Energy Consumption by Sector; 2016 Data.

Regulations of GHG Emissions from Vehicles

Light-duty vehicles (cars, SUVs, vans, and pickup trucks) and medium- and heavy-duty vehicles (including buses, heavy trucks of all kinds, and on-road work vehicles) are collectively the largest emitters of GHGs other than power plants. Together, on-road motor vehicles accounted for about 22% of U.S. GHG emissions in 2012. Reducing GHG emissions from this source category was a key component of President Obama's Climate Action Plan.[17]

EPA began to promulgate GHG emission standards for on-road vehicles in 2010, using its authority under Section 202 of the Clean Air Act. GHG standards for light-duty vehicles first took effect for Model Year (MY) 2012. Allowable GHG emissions will be gradually reduced each year from MY2012 through MY2025. In MY2025, emissions from new vehicles must average about 50% less per mile than in MY2010. The standards for heavier-duty vehicles began to take effect in MY2014. They will require emission reductions of 6% to 23%, depending on the type of engine and vehicle, when fully implemented in MY2018. A second round of standards, to address MY2021 and later medium- and heavy-duty vehicles, was finalized on October 26, 2016.

[17] See Executive Office of the President, *The President's Climate Action Plan*, 2013, https://www.whitehouse.gov/sites/default/files/image/president27sclimateactionplan.pdf; and CRS Report R43120, *President Obama's Climate Action Plan*, coordinated by Jane A. Leggett.

Many GHG emission reduction programs (e.g., the Regional Greenhouse Gas Initiative)[18] and legislative proposals have often focused on CO_2 emissions from the electricity generation sector, due to the sector's GHG emission contribution and the relatively limited number of emission sources. In addition, electric power plants have been measuring and reporting CO_2 emissions to the EPA for multiple decades.

CO_2 Emissions from Electricity Generation

Figure 5 compares U.S. electricity generation with CO_2 emissions from the electricity sector between 1976 and 2016. As the figure illustrates, U.S. electricity generation generally increased between 1976 and 2007 and then decreased in 2008 and 2009. Historically, CO_2 emissions from electricity generation followed a similar course. However, in 2010, these trends decoupled. While electricity generation remained flat after 2010, CO_2 emissions continued a general trend of reduction. Thus in 2016, electricity generation was essentially equivalent to generation in 2005, while CO_2 emissions were 25% below 2005 levels.

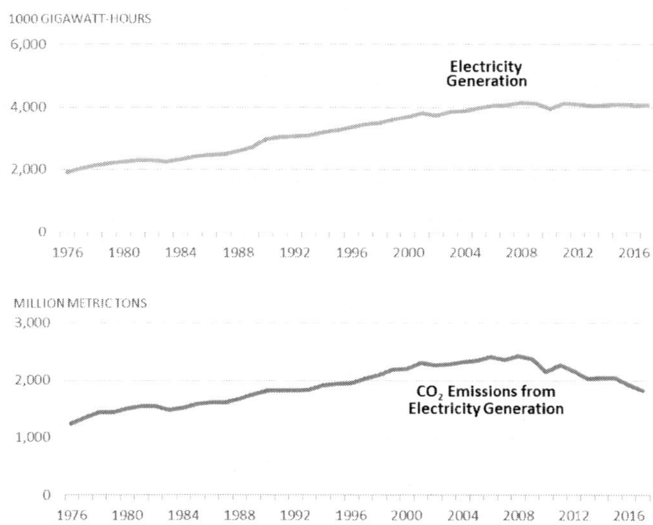

Source: Prepared by CRS; data from EIA, *Monthly Energy Review*, net electricity generation from Table 7.2 and emissions from Table 12.6, http://www.eia.gov/totalenergy/data/monthly/.

Figure 5. Electricity Generation and CO_2 Emissions from U.S. Electricity Sector; 1976-2016.

The decrease in CO_2 emissions in the electricity sector in recent years was likely a result of several factors. The economic downturn in 2008 and 2009 likely played a substantial role in both generation and emission levels. The U.S. gross domestic product

---

[18] See CRS Report R41836, *The Regional Greenhouse Gas Initiative: Lessons Learned and Issues for Congress*, by Jonathan L. Ramseur.

(GDP) decreased in both of those years. Historically, annual GDP decreases are a relatively uncommon occurrence: The United States has seen an annual decrease in GDP seven times over the last 50 years. The 2.9% GDP decrease in 2009 was the largest GDP decrease during that time frame.[19] Another factor contributing to the recent decrease in CO_2 emissions from electricity generation was the change in the electricity generation portfolio. Electricity is generated from a variety of sources in the United States. Some sources—nuclear, hydropower, and some renewables— directly produce no CO_2 emissions with their electricity generation. Fossil fuels generate different amounts of CO_2 emissions per unit of electricity generated. Figure 6 illustrates the relative comparison of CO_2 emissions between electricity produced from coal, petroleum, and natural gas. As the figure indicates, petroleum-fired electricity yields approximately 80% of the CO_2 emission of coal-fired electricity per kilowatt-hour of electricity.

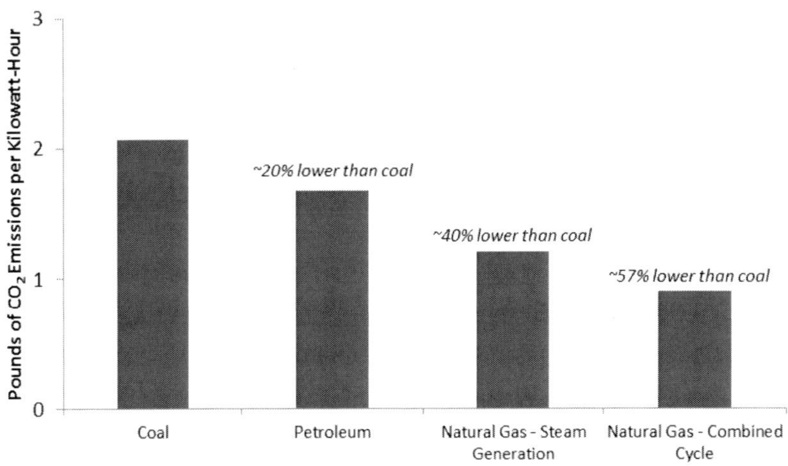

Source: Prepared by CRS; data from EIA, *How Much Carbon Dioxide Is Produced per Kilowatthour When Generating Electricity with Fossil Fuels?*, https://www.eia.gov/tools/faqs/faq.cfm?id=74 &t=11.

Notes: Carbon content values are derived by multiplying the fuel's CO2 emission factor by the heat rate of a particular electric generating unit. In this figure, CRS used the coal emission factor for bituminous coal and the petroleum emission factor measure for distillate oil (number 2). Natural gas has only one factor. The heat rates of different electricity unit types can vary substantially. CRS used EIA's average steam generation value for coal, petroleum, and natural gas, as well as the average combined cycle value for natural gas. The above comparison does not account for the so-called life-cycle emissions associated with the energy supply chain. For more information, see CRS Report R44090, *Life-Cycle Greenhouse Gas Assessment of Coal and Natural Gas in the Power Sector*, by Richard K. Lattanzio.

Figure 6. Comparison of Fossil Fuels' Carbon Content in Electricity Generation.

Natural-gas-fired electricity from a steam generation unit yields approximately 60% of the CO_2 emissions of coal-fired electricity per kilowatt-hour of electricity. Natural-gas-

[19] Bureau of Economic Analysis, gross domestic product data, http://www.bea.gov/national/index.htm.

fired electricity from a combined cycle unit yields approximately 43% of the CO_2 emissions of coal-fired electricity per kilowatt-hour of electricity.[20] Therefore, altering the U.S. electricity generation portfolio would likely have (all else being equal) a considerable impact on emissions from the electricity sector, which in turn, would have a meaningful impact on total U.S. GHG emissions. Figure 7 illustrates the percentage of electricity generated by source between 2005 and 2016. As the figure indicates, the U.S. electricity generation portfolio has changed considerably in recent years. Highlights include:

- Coal: Between 2005 and 2016, coal-fired generation decreased by 38%. Its contribution to total electricity generation decreased from 50% to 30%.
- Natural gas: Between 2005 and 2016, natural-gas-fired generation increased by 81%. Its contribution to total electricity generation increased from 19% to 34%. *In 2016, natural gas surpassed coal in terms of percentage of total generation.*
- Renewable energy: Between 2005 and 2016, non-hydro renewable energy generation increased by 254%. Its contribution to total electricity generation increased from 2% to 8%.
- Petroleum: Between 2005 and 2016, petroleum-fired generation decreased by 87%. Its contribution to total electricity generation decreased from 2% to less than 1%.

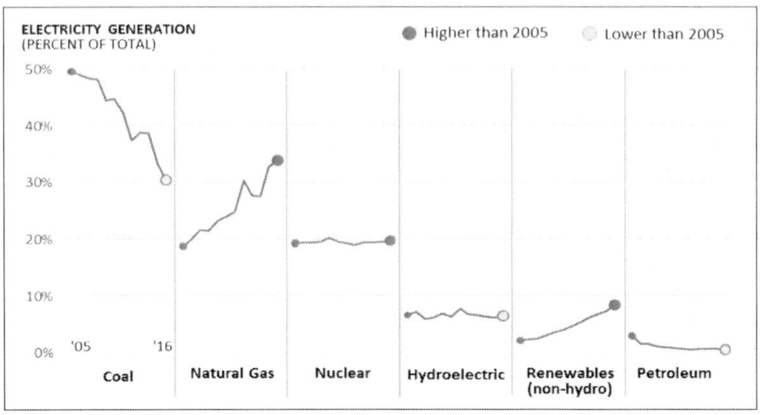

Source: Prepared by CRS; data from EIA, *Electric Power Monthly*, Table 1.1, http://www.eia.gov/beta/epm/.

Notes: Renewable sources include wind, utility scale solar, wood fuels, landfill gas, biogenic municipal solid waste, other biomass, and geothermal. Petroleum includes petroleum liquids and petroleum coke.

Figure 7. Percentage of Total Electricity Generation by Energy Source; 2005-2016.

[20] For further discussion, see CRS Report R44090, *Life-Cycle Greenhouse Gas Assessment of Coal and Natural Gas in the Power Sector*, by Richard K. Lattanzio.

Several factors likely played a role in these recent changes. Due in large part to technological advances—particularly directional drilling and hydraulic fracturing[21]—U.S. natural gas production increased dramatically (by 46%) between 2005 and 2016.[22] Natural gas production levels reached record levels in 2011 and have increased further each year until declining in 2016.[23] Relatedly, the weighted average annual price of natural gas dropped by about 60% between 2005 and 2015. By comparison, the weighted average annual coal price increased by about 40% during that time frame.[24] This change in relative fuel prices has played a key role in altering the economics of power generation (i.e., order of dispatch), leading to some natural gas displacement of coal in particular regions of the country.[25]

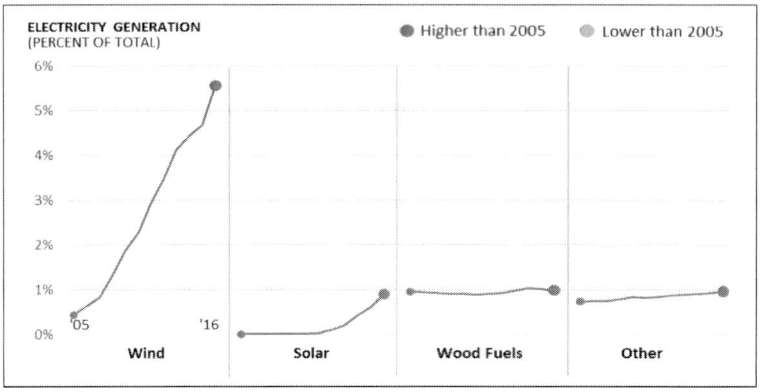

Source: Prepared by CRS; data from EIA, *Electric Power Monthly*, Table 1.1A, http://www.eia.gov/beta/epm/.

Notes: Solar generation does not include estimates of distributed solar generation, which EIA began to provide in 2014. Including these estimates would increase the percentage of solar generation in 2014 from 0.4% to 0.7%, in 2015 from 0.6% to 1.0%, and in 2016 from 0.9% to 1.4%. The "other" category includes landfill gas, biogenic municipal solid waste, other biomass, and geothermal sources.

Figure 8. Percentage of Total Electricity Generation from Renewable Energy Sources (Not Including Hydroelectricity); 2005-2016.

Figure 8 provides a more detailed breakdown of the changes in generation from non-hydro renewable energy sources. The vast majority of the increased generation from renewable energy over the past 11 years is due to wind power, which increased 13-fold

[21] Hydraulic fracturing is an industry technique that uses water, sand, and chemicals under pressure to enhance the recovering of natural gas and oil. It has taken on new prominence as it has been applied to tight oil and shale gas formation as an essential method for producing resources from those types of formations. See CRS Report R43148, *An Overview of Unconventional Oil and Natural Gas: Resources and Federal Actions*, by Michael Ratner and Mary Tiemann.

[22] EIA, "U.S. Dry Natural Gas Production, 1930-2016," http://www.eia.gov/dnav/ng/hist/n9070us2a.htm.

[23] In 2016, natural gas production decreased by 2%, but remained above 2015 levels.

[24] EIA, *Electric Power Annual*, Table 7.4, https://www.eia.gov/electricity/annual/html/epa_07_04.html.

[25] See for example, EIA, *Fuel Competition in Power Generation and Elasticities of Substitution*, 2012, http://www.eia.gov/analysis/studies/fuelelasticities/.

between 2005 and 2016. Although solar increased 67-fold over that time frame, the magnitude of wind generation dwarfs solar generation (227 Terawatt-hours of wind in 2016 versus 37 Terawatt-hours of solar). Energy from wood fuels has remained relatively constant during this time frame. The increase in "other" renewable sources is due to increased use of landfill gas, which more than doubled between 2005 and 2016.

ELECTRICITY CO_2 EMISSION PROJECTIONS AND THE CPP

As previously discussed, CO_2 emissions from fossil fuel combustion account for the vast majority (77%) of total U.S. GHG emissions, and the electric power sector contributes a large percentage (35%) of CO_2 emissions from fossil fuel combustion. Therefore, policymakers and stakeholders are paying attention to both recent trends and future projections of CO_2 emissions in the electricity generation sector.

Multiple factors will likely impact CO_2 emission levels from the electricity sector. Some of these factors, identified below in no particular order, are interrelated:

- Economic impacts (e.g., level of GDP growth);
- Prices of fossil fuels—particularly natural gas—and renewable energy sources;
- Electricity generation portfolio (e.g., whether recent trends in coal, natural gas, and renewable energy use continue);
- Federal and/or state policy developments (e.g., CPP implementation, state renewable energy requirements); and
- Improvements in demand-side energy efficiency (e.g., commercial and residential electricity use).

Most, if not all, of these factors are difficult to forecast with precision. In particular, the fate of the CPP is unknown, which raises a question: What impact would the CPP have on CO_2 emissions in the electricity sector?

Although it has been widely reported that the rule would require a 32% reduction in CO_2 emissions from the electricity sector by 2030 compared to 2005 levels, this percentage reduction comes from EPA's modeling of the effects of the rule in conjunction with other factors. The rule would establish uniform national CO_2 emission performance rates—measured in pounds of CO_2 per megawatt-hour of electricity generation—and state-specific CO_2 emission rate and emission targets. Each state would determine which measure to use to be in compliance.

Figure 9 compares EPA's projections of CO_2 emissions in the electricity sector resulting from the final rule with historical CO_2 emissions (1990-2015) from the electricity sector and projected emissions under EPA's baseline scenario. This scenario does not include the 2015 renewable energy tax extensions, discussed below. The figure indicates that the final rule would reduce CO_2 emissions in the electricity sector by 32% in 2030 compared to 2005 levels. Under the baseline scenario (i.e., without the CPP), EPA projects a 16% reduction by 2030 compared to 2005 levels. In terms of metric tons, EPA's CPP scenario would reduce an additional 377 million metric tons (or 15 percentage points) of CO_2, compared to EPA's baseline scenario.

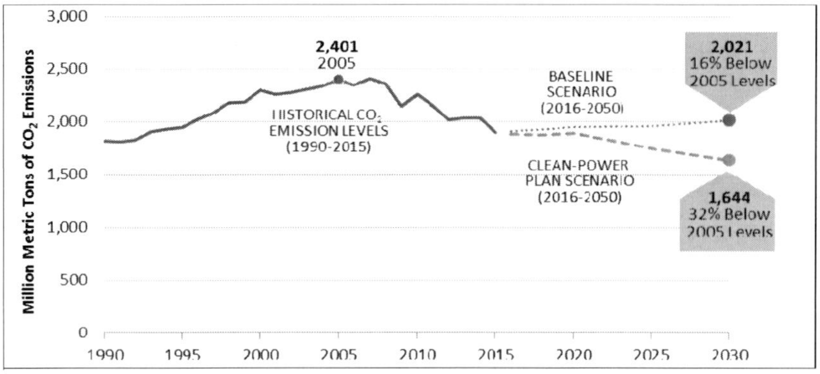

Source: Prepared by CRS; historical emissions from EPA, *Inventory of U.S. Greenhouse Gas Emissions and Sinks: 1990–2015*, April 2017; baseline and CPP projections from EPA, Power Sector Modeling, 2015, http://www.epa.gov/airmarkets/programs/ipm/cleanpowerplan.html.

Notes: CRS converted EPA's projected emissions from short tons to metric tons. The historical emission levels are included for illustrative purposes only. In its analysis, EPA compared its CO_2 emission projections to a 2005 emission level (2,433 million metric tons) produced from the Emissions & Generation Resource Integrated Database. This 2005 value includes emissions only from the contiguous United States; the historical levels in the figure include all 50 states. However, the CO_2emissions from Alaska's and Hawaii's electricity sectors generally account for less than 1% of total CO_2 emissions from that sector.

Figure 9. CO_2 Emissions from U.S. Electricity Generation; Historical Emissions, EPA Baseline Projection, and EPA Clean Power Plan Projection.

Based on EPA's *Regulatory Impact Analysis*, several factors may explain the emission results in 2030 between the baseline and CPP scenarios:

- Demand-side energy efficiency (DSEE) improvements: States may employ DSEE improvement activities as part of their plans to meet their targets. In its analysis, EPA assumed that DSEE efforts would decrease total electricity generation by 8% in 2030 compared to its baseline projection.[26]
- Coal generation decreases: EPA projected coal generation to decrease by 22% 23% in 2030 compared to its 2030 baseline projection.[27]

---

[26] EPA, *Regulatory Impact Analysis for the Clean Power Plan Final Rule*, August 2015 (hereinafter RIA), http://www2.epa.gov/cleanpowerplan/clean-power-plan-final-rule-regulatory-impact-analysis.

[27] EPA, RIA, Table 3-11. The range reflects the different compliance options in EPA's model: rate-based and mass-based approaches.

- Natural gas generation increases: EPA projected natural gas generation to increase by 5%-18% in 2030 compared to its 2030 baseline projection.[28]
- Renewable energy generation increases: EPA projected non-hydro renewable energy generation is expected to increase by 8%-9% in 2030 compared to its 2030 baseline projection.[29]

Other organizations used models to compare baseline scenarios with various CPP scenarios. Table 1 lists the CO_2 emission projections from these groups and EPA. Some of these groups produced multiple projections, employing different assumptions of future activities: CPP implementation options (e.g., whether states engaged in emissions trading) and levels of energy efficiency improvements, among others.

Table 1. Comparison of selected modeling projections: CPP and non-CPP scenarios million metric tons of CO_2 emissions

Modeling Group	**Non-CPP Scenario: 2030 CO_2 Emissions**	**% Below 2005 Levels**	**CPP Scenario(s): 2030 CO_2 Emissions**	**% Below 2005 Levels**
EPA (2015)	2,021	16%	1,644	32%
Energy Information Administration (2017)	1,886	22%	1,537	36%
Rhodium Group (2017)	1,774	26%	1,524	37%
M.J. Bradley and Associates (2016)	1,780-1,876	22%-26%	1,577-1,729	28%-34%
National Renewable Energy Laboratory (2016)	Not included	Not included	1,448–1,556	32%-36%

Source: EPA data from the agency's Power Sector Modeling, 2015, http://www.epa.gov/airmarkets/programs/ipm/cleanpowerplan.html; Energy Information Administration data from *Annual Energy Outlook 2017*, 2017, https://www.eia.gov/outlooks/aeo/. Rhodium Group data from "Taking Stock 2017: Adjusting Expectations for US GHG Emissions," 2017, http://rhg.com/reports/taking-stock-2017-adjusting-expectations-for-us-ghg-emissions; and personal correspondence with authors to provide 2030 estimate for CCP scenario. M. J. Bradley and Associates data from "EPA's Clean Power Plan Summary of IPM Modeling Results with ITC/PTC Extension," 2016, http://www.mjbradley.com/reports/updated-modeling-analysis-epas-clean-power-plan. National Renewable Energy Laboratory data from *Impacts of Federal Tax Credit Extensions on Renewable Deployment and Power Sector Emissions*, 2016, http://www.nrel.gov/docs/fy16osti/65571.pdf (and personal correspondence with report authors).

Notes: The groups in the table used different values for 2005 emission levels, but the differences were minimal. The percentage reductions in the table are based on the specific group's emission level in 2005.

All of the modeling scenarios below (except for EPA) included the December 2015 renewable energy tax extensions. On December 18, 2015, President Obama signed into law the Consolidated Appropriations Act, 2016 (P.L. 114-113). The act, among other

[28] Ibid.

[29] Ibid.

provisions, extended and modified the production tax credit (PTC) and the investment tax credit (ITC) for specific renewable energy technologies.[30] Prior to the December 2015 development, the PTC had expired and the ITC was scheduled to expire at the end of 2016. The PTC will not be available to projects starting construction after December 31, 2019. However, PTC tax expenditures will continue after that date, because the PTC is available for the first 10 years of renewable electricity production. The ITC for solar is scheduled to decline from 30% to 26% in 2020, and 22% in 2021, before returning to the permanent rate of 10% after 2021.[31]

In general, the modeling results in Table 1 indicate that the CPP would have a substantial impact on future CO_2 emission levels from electricity generation compared to scenarios that do not include the CPP.

CONCLUSION

Recent international negotiations and domestic policy developments have increased interest in current and projected U.S. GHG emission levels. In the context of international climate change negotiations, President Obama announced, on separate occasions, U.S. GHG emission reduction goals for both 2020 and 2025. Whether the United States ultimately achieves its goals would likely depend, to some degree, on CO_2 emissions from power plants.

Historically, CO_2 emissions from electricity generation have followed an upward course similar to electricity generation levels. However, in 2010, their courses diverged. While electricity generation remained flat after 2010, CO_2 emissions continued a trend of reduction. Thus in 2016, electricity generation was essentially equivalent to generation in 2005, while CO_2 emissions were 25% below 2005 levels.

Multiple factors generally impact CO_2 emission levels from the electric power sector. Recent changes in the U.S. electricity generation portfolio between 2005 and 2016 played a key role:

- Coal's contribution to total electricity generation decreased from 50% to 30%;
- Natural gas's contribution to total electricity generation increased from 19% to 34%, surpassing coal in percentage of total generation in 2016; and
- Renewable energy's contribution to total electricity generation increased from 2% to 8%.

---

[30] See National Renewable Energy Laboratory, *Impacts of Federal Tax Credit Extensions on Renewable Deployment and Power Sector Emissions*, February 2016, http://www.nrel.gov/docs/fy16osti/65571.pdf.

[31] For further information, see CRS Report R44852, *The Value of Energy Tax Incentives for Different Types of Energy Resources: In Brief*, by Molly F. Sherlock.

If these recent trends in the electric power sector continue, CO_2 emissions in that sector may continue to decrease. Assuming this were to occur, some might question the importance of the CPP in terms of meeting U.S. GHG emission goals (e.g., 26%-28% below 2005 levels by 2025). However, modeling results cited above indicate that the CPP would have a substantial impact on future CO_2 emission levels from the electric power sector.

INDEX

A

B

C

D

P

R

S

T

V

W

Airborne Particles: Origin, Emissions and Health Impacts

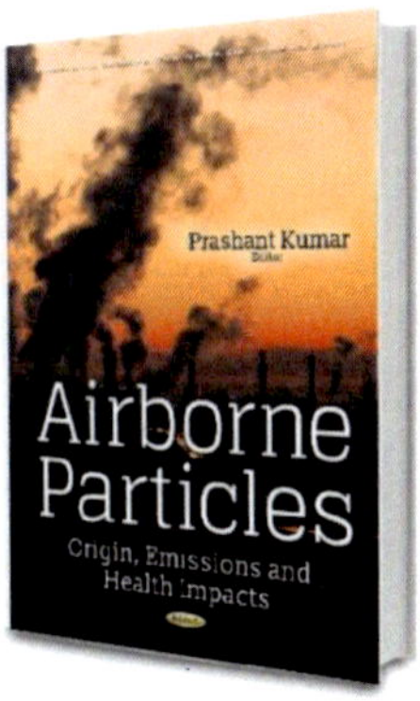

Editor: Prashant Kumar (Department of Civil and Environmental Engineering, Faculty of Engineering and Physical Sciences (FEPS), University of Surrey, Guildford, GU2 7XH, UK; Environmental Flow (EnFlo) Research Centre, Faculty of Engineering and Physical Sciences, University of Surrey, Guildford, GU2 7XH, UK)

Series: Environmental Remediation Technologies, Regulations and Safety

Book Description: *Airborne Particles: Origin, Emissions and Health Impacts* compiles a number of well-known authors in their respective research fields who have contributed their chapters on numerous specialised topics, such as sources of particulate matter emissions, their dispersion modelling, long-range transport, and both epidemiological and toxicological effects on human health.

Hardcover ISBN: 978-1-53610-965-8
Retail Price: $195

Environmental Variations in the North Atlantic in Response to Quaternary Glaciations

Author: Gloria M. Martin-Garcia, Ph.D. (Department of Geology, University of Salamanca, Salamanca, Spain)

Series: Climate Change and its Causes, Effects and Prediction

Book Description: In this book, the sea surface and deep water conditions in the mid-latitude North Atlantic during the Quaternary are reconstructed based on the analysis of both planktonic and benthic microfaunal assemblages from the SW Iberian margin.

eBook ISBN: 978-1-53613-367-7
Retail Price: $82

Environmental Change and Sustainability: Management, Challenges and Research Opportunities

Editor: Lydia Henderson

Series: Environmental Research Advances

Book Description: *Environmental Change and Sustainability: Management, Challenges and Research Opportunities* examines a variety of underlying conceptions regarding sustainability, providing a number of different views on its causes and effects.

Softcover ISBN: 978-1-53613-376-9

Retail Price: $82

Groundwater Contamination: Performance, Limitations and Impacts

Editor: Anna L. Powell

Series: Water Resource Planning, Development and Management

Book Description: Groundwater contamination is nearly always the result of human activity. This book focuses on the performance, limitations and impacts of groundwater contamination.

Softcover ISBN: 978-1-53611-003-6

Retail Price: $82